Advances in Computing, Electrical, Electronics, Mechanical and Communication Sectors

Advances in Computing, Electrical, Electronics, Mechanical and Communication Sectors

Edited by
Dr. S. Kannadhasan
Dr. R. Nagarajan
Dr. W. Deva Priya
Dr. T. Srihari
Dr. R. Saravanakumar

CWP
Central West Publishing

A catalogue record for this book is available from the National Library of Australia

NATIONAL LIBRARY OF AUSTRALIA

ISBN (print): 978-1-922617-43-9

Preface

The book's goal is to provide a premier interdisciplinary platform for leading academic scientists, industry peers, research fellows, and students to share the most up-to-date research and information on recent innovations, practical challenges, and solutions in the fields of Electrical, Electronics, Mechanical, Computer, and Communication Engineering. Quality research will support future advancements in a variety of engineering and technology areas. The book will facilitate worldwide collaborations by offering networking possibilities, developing appropriate solutions for a variety of applications and contemporary technology, and assisting in the creation of sustainable development solutions. The Advances in Computing, Electrical, Electronics, and Communication Sector Book Series seeks to publish research on a variety of engineering and technology-related subjects. The disciplines of computer, electronics, mechanical, communication, and electrical engineering include a wide variety of interdisciplinary subjects, allowing for multidisciplinary study. As many disciplines have become an essential part of daily life, research in these areas continues to grow and become more vital. This book may be helpful for both students and academics due to the wide variety of subjects addressed.

About Editors

Dr. S. Kannadhasan is working as an Assistant Professor in the department of Electronics and Communication Engineering in Study World College of Engineering, Coimbatore, Tamilnadu, India. He is Completed the Ph.D. in the field of Smart Antenna for Anna University in 2022. He is Twelve years of teaching and research experience. He obtained his B.E in ECE from Sethu Institute of Technology, Kariapatti in 2009 and M.E in Communication Systems from Velammal College of Engineering and Technology, Madurai in 2013. He obtained his M.B.A in Human Resources Management from Tamilnadu Open University, Chennai. He has published around 50 papers in the reputed indexed international journals indexed by SCI, Scopus, Web of science, Major indexing and more than 193 papers presented/published in national, international journal and conferences. Besides he has contributed a book chapter also. He also serves as a board member, reviewer, speaker, session chair, advisory and technical committee of various colleges and conferences. He is also to attend the various workshop, seminar, conferences, faculty development programme, STTP and Online courses. His areas of interest are Smart Antennas, Digital Signal Processing, Wireless Communication, Wireless Networks, Embedded System, Network Security, Optical Communication, Microwave Antennas, Electromagnetic Compatability and Interference, Wireless Sensor Networks, Digital Image Processing, Satellite Communication, Cognitive Radio Design and Soft Computing techniques. He is Member of SMIEEE, ISTE, IEI, IETE, CSI, IAENG, SEEE, IEAE, INSC, IARDO, ISRPM, IACSIT, ICSES, SPG, SDIWC, IJSPR and EAI Community.

Dr. R. Nagarajan received his B.E. in Electrical and Electronics Engineering from Madurai Kamarajar University, Madurai, India, in 1997. He received his M.E. in Power Electronics and Drives from Anna University, Chennai, India, in 2008. He received his Ph.D in Electrical Engineering from Anna University, Chennai, India, in 2014. He has worked in the industry as an Electrical Engineer. He is currently working as Professor of Electrical and Electronics Engineering at Gnanamani College of Technology, Namakkal, Tamilnadu, India. He has published more than 70 papers in International Journals and Conferences. His research interest includes Power Electronics, Power

System, Communication Engineering, Network Security, Soft Computing Techniques, Cloud Computing, Big Data Analysis and Renewable Energy Sources.

Dr. W. Deva Priya Completed her B.E. in Electronics and Communication Engineering from Madurai Kamarajar University, Madurai, Tamil Nadu, India, in 2004. She received her M.E. in Computer Science Engineering at Thaigarajar College of Engineering, Anna University, Chennai, Tamil Nadu, India, in 2008. She was awarded her Ph.D. in Information and Communication Engineering from Anna University, Chennai, Tamil Nadu, India, in 2017. She is currently working as Professor in Electronics and Communication Engineering at K S R Institute for Engineering and Technology, Tiruchengode, Namakkal, Tamil Nadu, India. She has more than 13 years of teaching Experience with Research and two years of Industrial Experience. She has published more than 30 papers in various Journals and Conferences and four patents. She received the Best Mentor Award from Texas Instrumentation, Best Project Award, and Best Innovates Practise Award from various agencies. She received Rs.22 Lakhs of funds from AICTE and 600$ worth of project kits from Texas Instrumentation for the student project contest. Her research interests are Image Processing, Advanced Driver Assistance System (ADAS), Deep Learning, Internet of Things, Communication Engineering and Design Thinking.

Dr. T. Srihari received his Bachelor's Degree in Electrical and Electronics Engineering from Thaigarajar College of Engineering, Madurai Kamarajar University, Madurai, Tamil Nadu, India, in 2001. He received his Master's Degree of Engineering in Power Electronics and Drives at PSG College of Technology under Anna University, Chennai, Tamil Nadu, India, in 2004. He was awarded a Ph.D. in Electrical Engineering from Anna University, Chennai, Tamil Nadu, India, in 2017. He is currently working as Professor in Electrical and Electronics Engineering at K S R Institute for Engineering and Technology, Tiruchengode, Namakkal, Tamil Nadu, India. He has more than 16 years of Teaching Experience and ten years of Research Experience. He has received a fund of Rs 25 Lakhs under various schemes and projects. He has published more than 45 papers in various Journals and Conferences and six Patents. His research interests are Power Electronics and Drives, Switched Reluctance Motor, Internet of Things, Image Processing, ADAS, Artificial Intelligence, Innovation, and Design Thinking.

Dr. R. Saravanakumar completed his B.E degree in Electronics and Communication Engineering in the year 2005 from AlagappaChettiarGovernment College of Engineering and Technology, Karaikudi and subsequently he completed M.E. degree in Communication Systems at PSG college of Technology, Coimbatore. He completed his Ph.D. in the area of Antenna Design and Wireless Communication from Anna University, Chennai in the year 2019. He has 12 years of Teaching Experience. Currently he is working as associate professor in the Department of Electronics and Communication Engineering, Saveetha School of Engineering. His area of interest includes monopole and dipole antenna design, micro strip resonator, wireless communication, wireless sensor networks, RF and microwave systems. His patent granted from IP Australian Government and one more patent published in Indian Patent. He has published 10 International Journal papers and20 International/National Conference papers. In addition, he has published one book, in the title of Electromagnetic Fields. He is a reviewer of AdHoC &Sensor Wireless Networks. He is lifetime member of various professional bodies like ISTE, IEI, IFERP and IAENG. He received funds from Central Government and State Government, Science and Technology departments.

Table of Contents

Chapter 1

DETECTION AND MITIGATION OF DISTRIBUTED DENIAL OF SERVICE (DDos) ATTACK USING SOFTWARE DEFINED NETWORK (SDN)

I. Varalakshmi, Department of CSE, Manakula Vinayagar Institute of Technology, Puducherry, India
M. Thenmozhi, Department of CSE, Pondicherry Engineering College, Puducherry, India

ABSTRACT

IoT refers to the Internet of Things, which is an emerging technology in this modern digital era. IoT establishes the communication between end devices using the internet with less human intervention. The data are collected from various sensors and actuators in the preprocessing layer of IoT. The network and transport layer are integrated and called as communication layer. The impact of vulnerabilities, malwares and attacks are increasing dramatically in the communication layer. This chapter deals with the security impacts and challenges in the communication layer of IoT. The exponential growth of Internet of Things (IoT) devices with limited computing resources and poor security configurations make the network more vulnerable. The malware attack which is massive in troubling the traffic in the internet is caused by Distributed Denial of Service (DDoS) attack, Spam attack, Ransomware attack, Cloud attack etc affect the regular flow of traffic and make the victim devices crash/shutdown. There exist so many solutions obtained for detection and mitigation of DDoS attack using Machine learning, Deep Learning and produces high accuracy in detection. This proposed chapter provides solutions for DDoS attacks using Block chain, IoT technology and SDN architecture. The detection of Distributed Denial of Service attacks based on Entropy method is proposed with flow-based analysis using Software Defined Architecture (SDN). The detection of DDoS is done based on the calculation of traffic flow, packet count with time interval and accuracy. The reduced threshold value, traffic pattern and less delay time improves the detection accuracy. Block chain is a system of recording information in a decentralized environment that is almost unfeasible to append data in a dataset. Mitigation of DDoS attack is based on public and distributed infrastructure to advertise Genuine, Suspected, and Malicious networks using Software Defined- Ethereum Virtual Machine (SD-EVM). The Malicious packets are mitigated and removed from the network and secure the network using blocks. The various classifier algorithms for detection and mitigation techniques are discussed in this chapter along with parameters and results obtained.

1. INTRODUCTION

The Internet of Things (IoT) broadly refers to the integration of physical devices, sensors and actuators. Internet of Things (IoT) an emerging technology in the digital era, all IoT devices are connected via the Internet. There are around 1.3 billion devices connected at the end of 2019 and will reach 17.5 billion IoT devices in 2025. IoT enables people to live and work smarter and also enables the companies/ organizations and devices to automate processes and reduce labor cost. In most Internet scenarios, devices interact with applications that run remotely on the network, which enables malicious agents to take control of devices; here security is a big challenge.

IoT consists of five layer and three layer models. The important layers are Perception layer, Communication Layer and Application layer. IoT can operate, actuate, and communicate autonomously to optimize and enable new services in a wide range. The data are collected from various sensors and actuators. Machine to Machine Communication refers to exchanging of data between various devices through the internet without human intervention. The M2M first collects data from various sensors and actuators that transmit the data/information through a wireless network using the Internet and processing the data using algorithms.

Cyber attacks are the major issues in Internet of Things (IoT) worldwide. In 2025, as per the world market survey the number of active IoT devices would be increased by 75 billion. 98% accuracy achieved in existing systems using Machine learning algorithms using few parameters. DDos attack is detected in an online stream monitoring system using MQTT protocol in the network layer.

2

MQTT (formerly referred to as MQ Telemetry Transport). The MQTT protocol depends on the publish/subscribe communication pattern and uses a broker to facilitate message exchange between end-points. In an IoT scenario, the publishers are various low power sensors publishing data like temperature, pressure, humidity and subscribers could be smart devices like home automation controller, smart meter, Smartphone, computer who can take actions based on the environment changes.

1.1 SECURITY THREATS AND VULNERABILITIES IN IoT

IoT has developed a high degree of connectivity, customization and automation. The internet is a complex network connected by millions and billions of servers and systems through a global network that allows storage, retrieval and sharing information between devices. Cybercriminals are constantly searching for vulnerabilities to steal information, extort business and take control of systems remotely. The following are the top IoT vulnerabilities identified and having the biggest impact on users:

- **Weak, guessable passwords:** Most of the IoT devices come with preset credentials by the manufacturer. These credentials are available publicly and can be easily broken through brute force attacks.
- **Insecure Network services-** The devices are to connect with the end point, a simplest attack is tried to find the weakness in machine communication and services running on the device.
- **Unhealthy IoT ecosystem –** Users may accidently introduce security vulnerabilities at the application layer. This includes weak authentication controls, encryption protocols and unoptimized IO filtering.
- **Inefficient update mechanisms:** The IoT devices are updated each time once get activated. Outdated versions could be glaring with code vulnerabilities.
- **Botnet:** It is a network that collects devices together to remotely have control over a target system and distribute malware. Mirai botnet is a dangerous IoT security threat so far. It steals confidential data, and executes cyber attacks like DDos and phishing.
- **Ransomware**: This becomes the most tarnished cyber attack. Hackers use malware to encrypt data and decrypt vital data only after receiving a ransom. It can also be used to attack IIoT devices and smart homes.

3

- **Improper data transfer and storage:** Attackers gain access to an IoT device, ahead control over it and being able to alter the data it collects.

1.2 DDoS ATTACK

Distributed denial-of-service (DDoS) attack is one of the most challenging attacks for the researchers. It involves flooding the victim systems internet traffic so that it is unusable. It evolves into a serious threat to the flow level of both business and government organizations. The goal is to make the website or server crash/shutdown. This attack utilizes many compromised systems/devices as source of the traffic which bot accessed through this achieves effectiveness. The major threats in DDoS attacks: Ransomware attacks in the form of malware that encrypts the victim's server by demanding money through some links, email links or instant messages. Mirai attack: a malicious code that replicates itself by finding the victim and infecting the target by injecting vulnerabilities. The flooding of requests to the victim's server and make them shutdown or block normal activity of the server. These are the two main DDoS attacks that are challenging for the researchers in IoT so far. Because the devices connected in the IoT network are unpredictable.

DDoS is mastered through a network through remote controlled, hacked computers or bots. These are often referred to as "zombies". Botnets can range from thousands to billions of computers monitored and controlled by cybercriminals. Cybercriminals use botnets for a spread of purposes, including sending spam and sorts of malware like ransomware.

1.3 EVOLUTION OF DDoS

DDos attacks are like "in the wild". The most dangerous and challenging threat in IoT is DDos attacks. This challenging attack is defined by the volume of packets/data hacked by the attacker. Mirai botnet is the largest volume threat attack other than DDos attacks that have attacked the country so far.

In the recent DDoS attack, Mirai Botnet in the year 2019 September, targets the Europe server and this technique is called Connectionless Lightweight Directory Access Protocol (CLDAP). It was sent 50-70 times with vulnerable third party CLDAP servers. It lasts three

4

days and causes AWD revenue losses. It attacks over 809 million packets per second.

2. RELATED WORKS

The low-rate DDoS attack against the data link layer was detected and the effectiveness of DDoS attack detection predicted using FM algorithm. The defense method based on dynamic deletion of flow rules, and carried out experimental simulation and analysis proved the effectiveness of the defense method, and produced the success rate of forwarding normal packets of 97.85 percent. The proposed structure is built with a C-DAD (Counter-based DDoS Attack detection) framework in the top of the SDN WISE framework to detect and analyze DDoS attack .In this framework they have controllers for the SDNWISE and IoT. The counter-based function has flow monitor and flow analyzes to detect the DDoS attack in the SDN-IoT network. The C-DAD algorithm and framework are tested with different parameters and the detection time was high. It proposed the SDN-based cross domain attack detection. They have used KNN to detect in low time periods and to get high accuracy. The precision value is high .9sec; the attack was not detected at the earlier stage, since the possibility of flooding of bot into the network is high. KFNN technique is proposed to detect the DDoS attack in SDN, to improve high efficiency and accuracy from the enhancement of KNN .Here high efficiency and accuracy produced for limited network and the analysis gets reduced if the number of resources increased. In proposed LEDEM leverages for detecting DDoS attacks that run on centralized cloud SDN architecture. The DDoS detection accuracy results were tested with benchmark dataset UNB-ISCX from the LEDEM test bed in SDN emulator. The detection accuracy is measured based on true positive and false negative for DDoS, false positive- true negative for benign node detection packets. SD-IoT is composed of an SD-IoT controller pool with controllers, SD-IoT switches integrated with the IoT gateway and terminal IoT devices. It can detect DDoS attacks but cannot locate the specific IP address. Ethereum, the time required to mine a new block is theoretically 14 seconds, for a victim under an attack waiting for 14 seconds to obtain validation about Collaborative DDoS

3. ALGORITHMS USED FOR DDoS DETECTION

3.1 ENTROPY BASED DETECTION ALGORITHM

The input given corresponds to a new packet that has arrived with a new source address. The destination IP address is also examined to see if it has an existing instance in the window. If it does exist, the count for that IP address will be incremented. If the window gets full, the entropy is computed and then compared with the threshold. If the computed value for an entry is higher than the threshold for five consecutive counts, it will be classified as an attack.

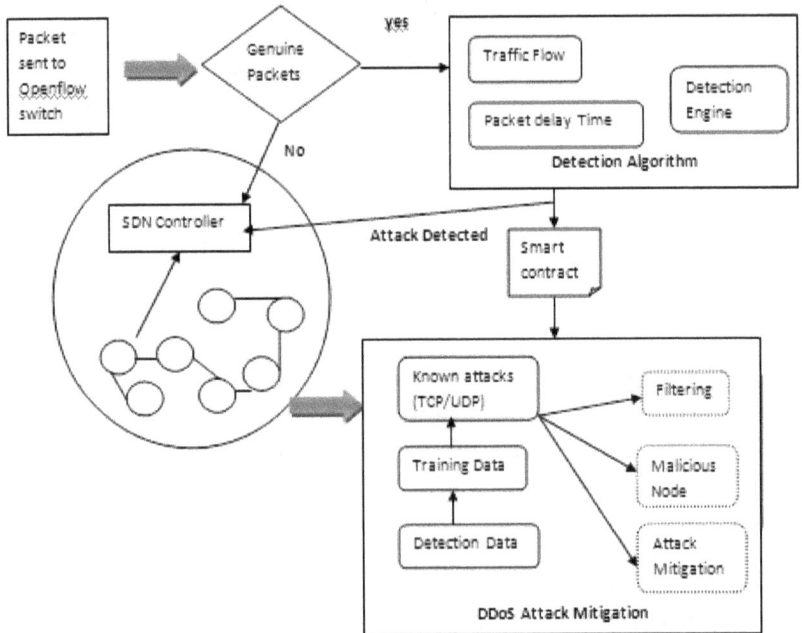

Fig: 1. Detection and Mitigation using SDN Architecture

The nodes are connected using TCP protocol and SDN enables the network behavior and centrally controlled manner through software applications using open APIs. SDN controller maintains source IP, Destination IP, Source and Destination port, packet arrival time are stored. The detection algorithm detects the malicious packets based on the training datasets. The malicious packets are sent to the filtering technique as shown in Fig.1. It filters the malicious and non malicious packets separately using classifier technique. The

6

malicious packets are sent to smart contracts and mitigate using blocks.

3.2 NAIVE BAYES CLASSIFIER

NB classification aims to detect the category, namely the category, of the info entered into the system with a series of calculations defined in accordance with the principles of probability. In NB classification, a certain amount of training data is entered into the system. The training data is entered and must have a class. Test data entered into the system with probability processes carried out on the training data is processed. This process is done according to previously obtained probability values and then the class of the given test data is detected. The more the amount of coaching data, the more accurate it's to detect the important category of the test data.

3.3 K-NEAREST NEIGHBORS

The KNN algorithm is a simple, easily applicable, and supervised machine learning algorithm that can be used for solving both classification and regression problems. When new data comes in, it determines the class of the new data by looking at its nearest K neighbors. Manhattan, Minkowski, and Euclidean distance functions are used for the distance between two data. The Euclidean distance function was used in this study. The similarity between the sample to be classified and the samples found in the classes was detected. When new data was encountered, the distance of this data to the data in the training set was calculated individually by using the Euclidean function. Then, the classification set was created by selecting the k dataset from the smallest distance. The number of neighboring KNN (k) is based on the value of classification. During the classification, k was determined as 10.

4. TECHNIQUES AND PROTOCOLS FOR MITIGATION

Mitigation refers to protecting the network from attacks, threats and vulnerabilities before floods. The mitigation mechanism protects end-users when a malicious attack is detected. To mitigate the attack, new flow rules are added to the switches with high priority to match the rule with the suspicious package. The precision of the mitigation rules depends on the amount of information the attack can be acquired. The controller configures the rules in the switches using OpenFlow standard messages, sending messages FlowMod. There are different mechanisms to mitigate the attack like black holes, the Intrusion Detection System (IDS), even advanced techniques such as Deep Packet Inspection (DPI). In this work, the mitigation algorithm is implemented, configuring different rules in the switch, which drops the suspicious packets.

4.1. FILTERING

DDoS Traffic is screened by identifying patterns that differentiates between legitimate traffic and hostile visitors instantaneously. Responsiveness is a way in which it is being able to block an attack without interfering with the user's experience. The main objective of the solution is that it should be completely transparent to the site visitors. One of the best mitigation strategies is to filter requests upstream known as Upstream Filtering. It is done effectively since the API never sees the traffic and so any rate limiting policies are not triggered. There are providers of "Mitigation Centers" that will filter the approaching network traffic. For example Amazon Shield and Cloudflare opens a new window and both offer products that allow for protection against DoS and DDoS attacks by checking incoming packet IPs against known attackers and BotNets and attempting to only forward legitimate traffic. Various API gateways have the same capabilities but can also filter based on the requested endpoint, and it allows HTTP verbs, or even a combination of verbs and endpoints

4.2 PORT MITIGATION TECHNIQUE

In order to secure the traffic flow, port mitigation techniques are applied. It collects malicious DDoS attack data from the detection system (Port ID, Destination IP, Flow). Prevent the network by analyzing the trained and test data available in the dataset associated with attack location. So that the attacker may not modify the traffic flow and execute the attack in the flow.

4.3 SMART CONTRACT

In a smart contract the packets and nodes have to register itself in another smart contract registry and store all related packets and

8

monitored. The autonomous systems listen for the changes in network and inclusion of nodes also to be monitored and assessed against the properties of the Autonomous system and apply mitigating rules wherever necessary. The Ethereum Virtual Machine (EVM) smart contracts sustain in a decentralized manner and local way the logic to control.

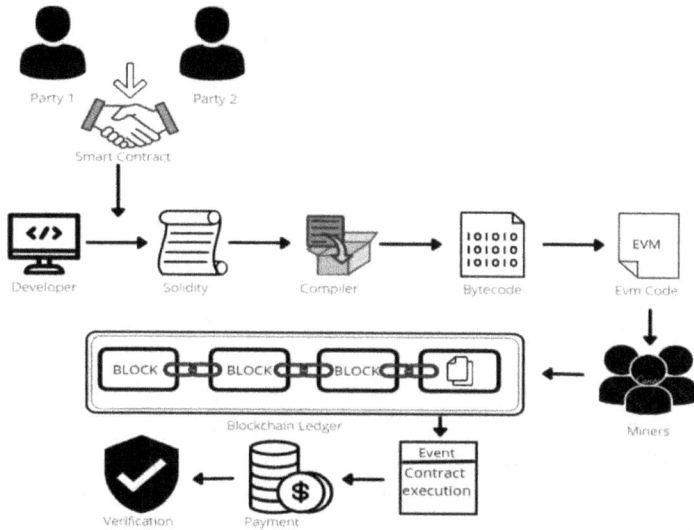

Fig: 2. Smart contract using blockchain

The detailed view of smart contracts in blockchain is shown in Fig.2. Malicious packets received from the detection algorithm are input to the smart contract. Smart contract checks the IP with trained dataset and EVM collects the malicious packets and stores them in datasets. From the miner, the packets are sent to blocks and secure the network from attacks.

5. CONCLUSION

In this proposed method, SDN architecture is being used for detection of DDoS attacks. Existing method produced an accuracy rate 85% with the flow analysis and detection algorithm method. In this approach, since the threshold value is updated adaptively based on traffic pattern condition, the accuracy of detection is improved. After detecting DDoS attacks, as a future work it is possible to find out attackers using IP trace back mechanism, as the detection system can perform efficiently based on flow aggregation method. Our approach depends on IoT end devices threshold value in mitigating the DDoS attack to increase the detection accuracy and sensitivity. The algorithms are designed to be very lightweight to meet the IoT and blockchain requirements in terms of time, power, and memory consumption. The proposed algorithm also, can distinguish between Genuine, Suspected and Malicious DDoS attacks accurately and mitigate them very efficiently. According to the analysis, the techniques and algorithms are analyzed and effectively helps the IoT environment for detecting and mitigating DDoS attacks and secure using block chain technology.

REFERENCES

[1] Nagarathna Ravi, Student Member, IEEE, and S. Mercy Shalinie, Senior Member, IEEE(2020) "Learning-Driven Detection and Mitigation of DDoS Attack in IoT via SDN-Cloud Architecture" *IEEE Internet of Things Journal*, Vol. 7, No. 4, April, ISSN No:2327-4662

[2] Shi-Ming Xia; Shi-ZeGuo; Wei Bai; Jun-Yang Qiu; Hao Wei; Zhi-Song Pan(2019), "A New Smart Router-Throttling Method to Mitigate DDoS Attacks" *published in IEEE.*, December 2019

[3] Da Yin, Lianming Zhang, Kun Yang,(Senior Member, IEEE)(2018) "A DDoS Attack Detection and Mitigation With Software-Defined Internet of Things Framework" published in Security And Trusted Computing For Industrial Internet of Things, *IEEE*, December 2018.

[4] I.Varalakshmi, Sivaraj, Ajeethkumar, Sharath,(2019) "Smart Dumpster Monitoring System Using Efficient Route Finding Algorithm" presented a paper in IEEE - International Conference on Systems Computation, Automation and Networking on 29th & 30th March 2019. *DOI: 10.1109/ICSCAN.2019.8878809*

[5] MeryamEssaid , DaeYong Kim , Soo HoonMaeng , Sejin Park, Hong TaekJu (2019) "A Collaborative DDoS Mitigation Solution Based on Ethereum Smart Contract and RNN-LSTM " *Asia-Pacific Network Operations and Management Symposium* (APNOMS).

[6] Wu Zhijun, Xu Qing, Wang Jingjie, Yue Meng and Liu Liang (2019),"Low-rate DDoS Attack Detection Based on Factorization Machine in Software Defined Network," *IEEE.* November 2019

[7] Jalal Bhayo, Sufian Hameed, and Syed Attique Shah (2020), "An Efficient Counter-Based Ddos Attack Detection Framework Leveraging Software Defined IoT (SD-IoT)," *IEEE.*, June 2020

[8] I.Varalakshmi, Pooja, Priya (2020), "Detection of Abnormal activities using Diffusion convolutional-Recurrent Neural Network" published in UGC approved "*International Journal of Management of Technology and Engineering (IJMTE)*" ISSN: 2249-7455 Vol 10 Issue 6, Page No: 247- 250. June 2020.

[9] Liehuang Zhu, Xiangyun Tang, Meng Shen, Xiaojiang Du and Guizani(2018), "Privacy-Preserving Ddos Attack Detection Using Cross-Domain Traffic in Software Defined Networks," *IEEE.* , July 2019.

[10] I.Varalakshmi, Farzana Begum, Saranya, Muthamizselvi(2019), "IoT Based Navigation System For The Visually Challenged" published in UGC Approved "*International Journal of Scientific Research and Review (IJSRR)*". ISSN NO: 2279-543X Vol 8, Issue 4

[11] Yuhua Xu, Houtao Sun, Feng Xiang and Zhixin Sun (2019), "Efficient DDoS Detection Based on K-FKNN in Software Defined Networks," *IEEE.*

[12] Tamer Omar, Anthony Ho, Brian Urbina (2019), "Detection of Ddos in SDN Environment Using Entropy-based Detection", *IEEE Conference on Technologies for Homeland Security.*, August 2019

[13] I.Varalakshmi (2016) "Botnet Detection using Iterative Filtering Algorithm in mobile Adhoc network" published in "*International Journal for Research in Applied Science & Engineering Technology (IJRASET)*", Vol 12, Issue 5, ISSN: 2321 9653

[14] I.Varalakshmi, S.Kumarakrishnan (2019), "Navigation System for the Visually Challenged Using Internet of Things" presented a paper in *IEEE - International Conference on Systems Computation, Automation and Networking. DOI:10.1109/ICSCAN.2019.8878809,* June 2019

[15] Jesús Galeano-Brajones, Javier Carmona-Murillo, Juan F. Valenzuela-Valdés and Francisco Luna-Valero (2019), "Detection and Mitigation of DoS and DDoS Attacks in IoT-Based Stateful SDN: An Experimental Approach". *IEEE,* December 2019

[16] Lohit Barki; Amrit Shidling; Nisharani Meti; D G Narayan; Mohammed Moin Mulla (2016) "Detection of Distributed Denial of Service Attacks using Machine Learning Algorithm in Software Defined Networks" *2016 International Conference on Advances in Computing, Communications and Informatics* (ICACCI) 10.1109/ICACCI.2016.7732445.

Chapter 2

SMART LINE FOLLOWER ROBOT WITH OBSTACLE DETECTION

S.Pragadeswaran[1], S.Gayathri[2], S.Gopinath[3]
Karpagam Institute of Technology[1], Karpagam
Institute of Technology[2], Karpagam Institute of
Technology[3], India

SYNOPSIS

Robots are usually designed for reducing human effort and making their human life get easier in life. Supported with the aim of human, many robots are designed for practical applications that it's going to be the simplest results obtained by only monitoring or manual operation. The automation line follower robot used predefined operation for transporting one place to a different place. Path is predefined. During line follower operation if any obstacles in path boot get damaged or work get stop due to obstacle the problem occurs. To avoid this problem we developed line follower with an obstacle avoid and that robot avoids the obstacle and gets back to the road. The obstacles are often avoided by using IR sensor. These types of robots can perform a lot of tasks in industrial works like material handling, hospitality works etc.

INTRODUCTION

Robots are usually designed for reducing the human efforts and making their lives easier to live. Based on the purpose of humans many robots are designed for periodical application. It may be any projects the best results are only obtaining by the continuous monitoring. Nowadays the industries are using the smart and intelligent live follower robot for the easy moving of goods from one place to another. The few reasons why robots can be employed for transportation of goods are for its fits and forget ability, first thing which done is the robots is placed in the desired path where coming to the working of the robot is totally automatic and this is the main reason for using robots is that robots used not to be manually controlled. This makes the follower robot more efficient than the conventional robots. There is a no particular path for a robot to travel and hence traditional obstacle avoiding robot will not be useful in transportation of goods because it will not reach to the required decision. The obstacle avoiding robot's mobility is uncontrollable. A Wi-Fi controlled robot is also not helpful because it needs manual application or operations. The application of these type robots are limited in workplaces because it can go in direction but the main problem is it needs continuous manual commands. Considering all these factor line follower robot has more useful application. By giving the conventional line follower to give the ability to detect obstacles can make them smart and intelligent. This improves the working of line follower robot, because in any working environment the obstacles are often and if the robot is unable to detect the obstacle they will collide and will be completely damaged. Then by adding the obstacle finding robots features to the line follower robots removes the damage to the robots. These robots can also be used in health care section in hospital for monitoring the patient reducing the human effort. By making use of these robots in transporting good the workers allocated for that work can be assigned to other works.

HARDWARE DESCRIPTION

2.1 IR sensors: The unphased sensor consists of Infrared LED and Infrared photo diodes. The IR LED is photo emitter and IR photo diode is the receiver. The light emitted by the LED gets deflected back to the photo diodes. The output voltage produced by the photo diode is proportional to the reluctance of the surface and low for dark surface which means that dark-colored objects reflects less IR lights than the light colored objects.

Figure 1 - IR Sensor

2.2 Ultrasonic Sensor: It can measure the distance of an object by using sound waves. It sends sound waves at a particular frequency to measure the distance, and it listens after the waves bounces back. Furthermore, it cannot detect an obstacle when the sound wave moves from its path, and it will not receive by the ultrasonic sensor, and it will not detect the obstacle. It also cannot receive the wave when the obstacle is too small. The ultrasonic sensor accuracy is depends upon the temperature and humidity of the area.

Figure 2 - Ultrasonic sensor

2.3 Ultrasonic Sensor: It can measure the distance of an object by using sound waves. It sends sound waves at a particular frequency to measure the distance, and it listens after the waves bounces back. Furthermore, it cannot detect an obstacle when the sound wave moves from its path, and it will not receive by the ultrasonic sensor, and it will not detect the obstacle. It also cannot receive the wave when the obstacle is too small. The ultrasonic sensor accuracy is depends upon the temperature and humidity of the area.

Figure 3 – Arduino Board

2.4 Motor Driver: It is used to control the current in the motor, and it also acts like a current amplifier. If the circuit receives the low current the motor driver would provide high current to the motor. These motors need high value of current to drive. To control two dc motors at the same time IC L293D will be used. The motor can run in both forward and reverse direction. When the robot needs to turn left or right the motors will be controlled by a motor driver. It completely controls the dc motors.

Figure 4 - Motor driver

WORKING PRINCIPLE

The line follower robot is made up of infrared sensor which is induced with LED and photo-diode. LED transmits light which strikes on the surface and it's reflected back to the photo-diode. Photo-diode absorbs the light which is reflected on it. The surface which is light (white surface) has high output and the surface which is dark (black surface) has low output. In this robot 2 IR sensors are used which makes the robot follow the black line. This line follower robot can be achieved by giving the output of the IR sensor to the Arduino controller. The Arduino coding part where the coding can be written based on the operation. For example if the output of the IR sensor is high, then the motor will run else it will be turned off. This logic is achieved using the Arduino controller. Motor driver used between the motor and Arduino and motor is used for the rotation of forward and reverse direction.

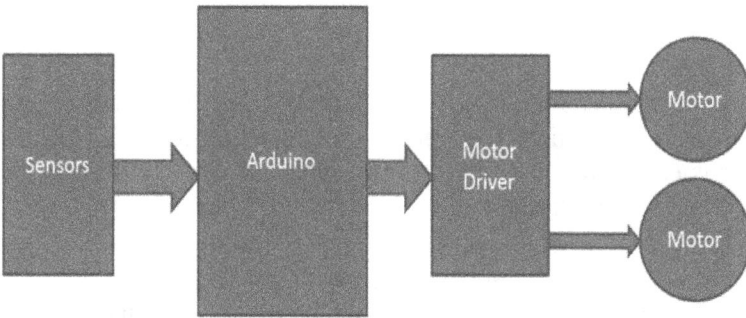

Figure 5 - Block diagram

If the output of the 2 IR sensors is high, then output of the motor is high i.e. the 2 motors will run in the forward direction. Even if anyone of the IR sensor has a low output the motor will not be able to run. Hence, using this concept the line follower works.

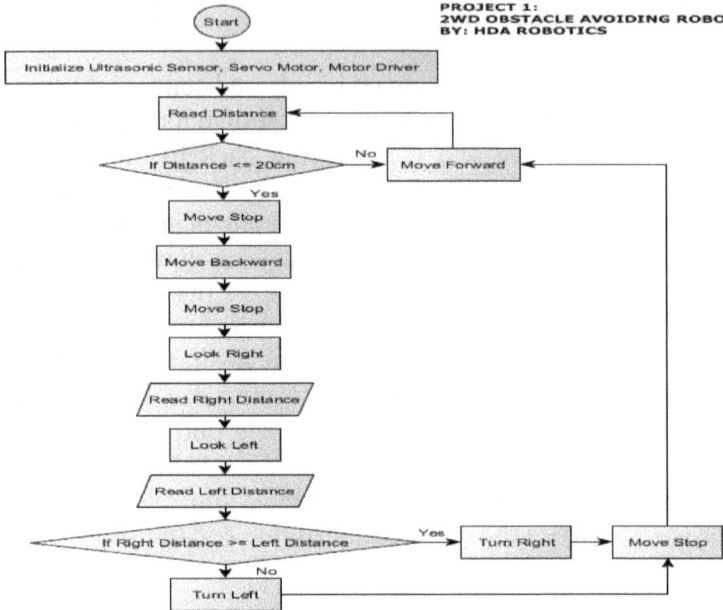

PROJECT 1:
2WD OBSTACLE AVOIDING ROBOT
BY: HDA ROBOTICS

Figure 6 - Flow chart

An obstacle avoider robot has the same concept as a line follower robot but with a small difference. In a line follower robot the IR sensor is connected at the bottom to follow the black line whereas in an obstacle avoider robot it is connected in the front for detecting if anything is present in the way at a particular distance. After identifying the obstacle the robots forwards to either right or left. Firstly the robot checks in the right if any obstacle is present then it moves to the left. This operation isperformed for the continuous movemet of the robot. If the output of the IR sensor is low it will move to the forward direction otherwise it will check right or left then moves in the forward direction.

18

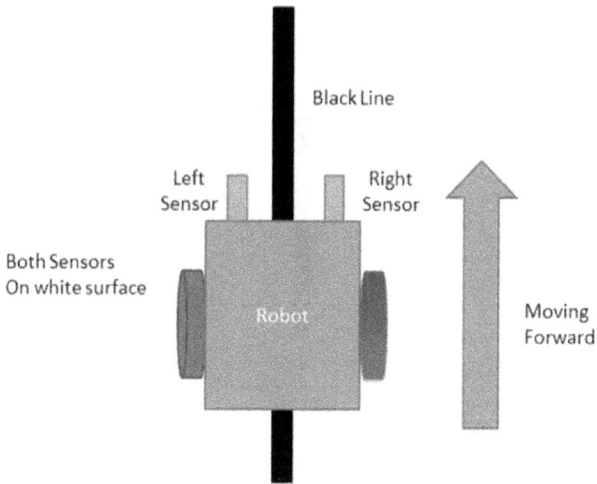

Figure 7 - Forward movement

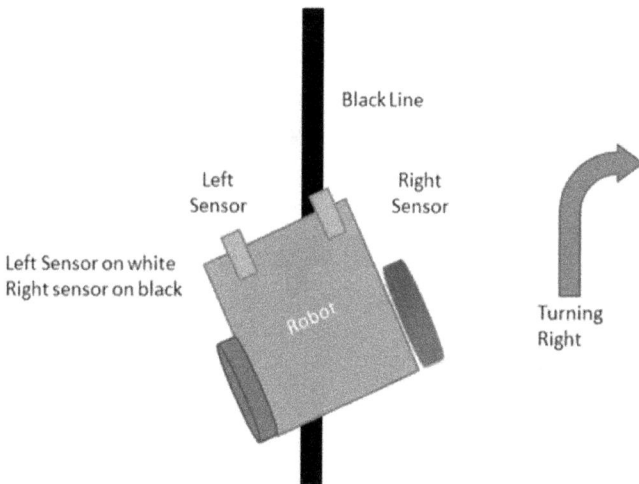

Figure 8 - Turning Right

This combinational robot has 5 IR sensors. The 1st and 2nd sensors are used to read the values for the line followed by the robot. The 3rd sensor is used in-front of the robot to detect the obstacle which is present in the line followed by the robot. The 4th and 5th are the left and right sensors which are used to overcome the obstacle and to retain back to the original line which should be followed by the robot to reach the destination.

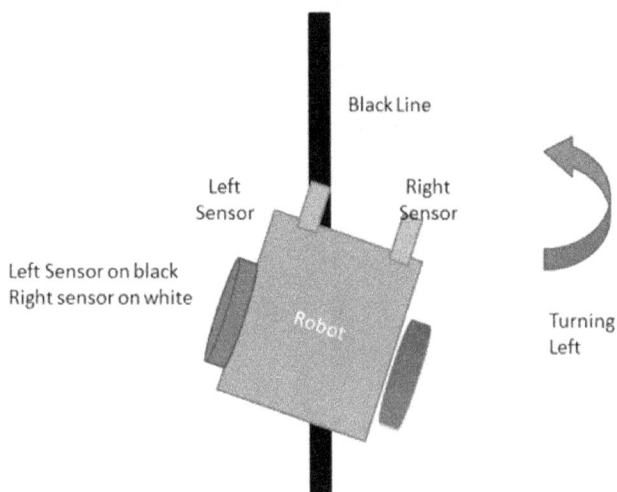

Figure 9 - Turning left

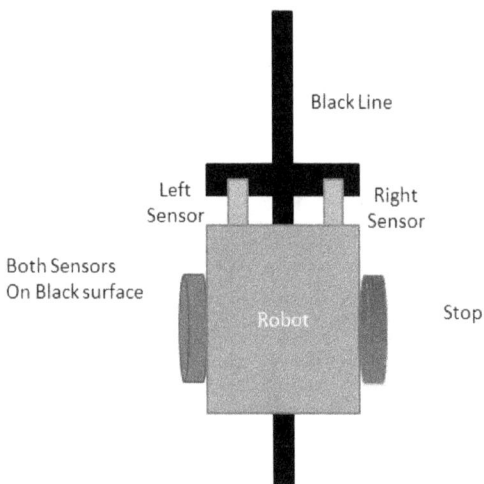

Figure 10 - Stop the robot

If any obstacle is sensed by the robot the right side sensor senses right side and if the sensor is low the robot takes 90 degree right side and it moves forward. The robot moves forward until the left side sensor becomes zero. If left side sensor is low the robot takes 90 degree left and moves forward. The robot moves forward until the left side sensor becomes zero. If left side sensor becomes zero the robot will takes 90 degree left side and the robot will moves forward. Now the robot reaches the black line and now the line senor becomes high and the obstacle sensor becomes zero. Now the robot takes 90 degree right side on the line and the robot will move on the line with line follower.

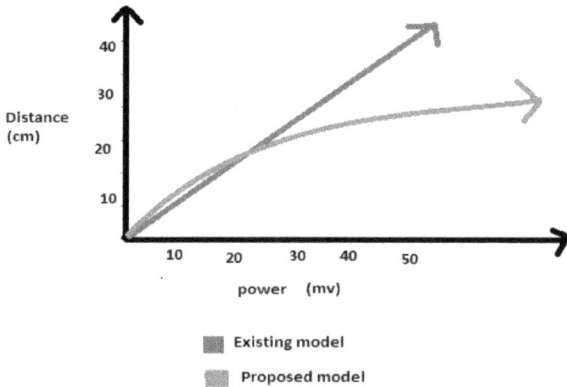

Figure 11- Graph

CONCLUSION AND FUTURE SCOPE

An obstacle avoiding the robot with the line follower identifies the object and avoiding the object with retaining the line which can be any controller. And an obstacle detect by either IR sensor or ultrasonic sensor. 5 IR sensors using for object detection. The robot has been fully autonomous and coding is given at initial stage. It can be got maximum output only by coding. Main advantage is very less cost with maximum performance. It Can be used various industrial purpose particular of automobiles. During pandemic period the robot can be use to providing food and medicine to patient.

REFERENCES

[1] Román Osorio, C., José Romero, A., Mario Peña, C., and Ismael López-Juárez, (2006) Intelligent Line Follower Mini-Robot System. *International Journal of Computers, Communications & Control*, **1**(2), 73-83.

[2] Deepak Punetha, Neeraj Kumar and Vartika Mehta, (2013) Development and Applications of Line Following Robot Based Health Care Management System. *International Journal of Advanced Research in Computer Engineering & Technology* (IJARCET), **2**(8).

[3] EndrowednesKuantama, Albert Brian Lewis Lukas and Pono Budi Mardjoko, (2014) Simple Delivery Robot System Based On Line Mapping Method. *ARPN Journal of Engineering and Applied Sciences*, **9**(11), 2078-2083.

[4] Faiza Tabassum, Susmita Lopa, Muhammad Masud Tarek and Bilkis Jamal Ferdosi, Obstacle Avoiding Robot (2015). *Global Journal of Researches in Engineering.*, Vol.12. Issue 12

[5] FaradilaNaim and Tan Piow Yon, (2011) Two Wheels Balancing Robot with Line Following Capability. *International Journal of Mechanical and Mechatronics Engineering*, **5**(7), 1401-1405.

[6] Rakesh Chandra Kumar, Md. Saddam Khan, Dinesh Kumar, Rajesh Birua, Sarmistha Mondal and ManasKr. Parai, (2013) Obstacle Avoiding Robot – A Promising One. *International Journal of Advanced Research in Electrical, Electronics and Instrumentation Engineering*, **2**(4), 1430-1434.

[7] Gadhvi Sonal Punit Raninga Hardik Patel, (2017) Design and implementation of RGB color line following robot, *International Conference on Computing Methodologies and Communication (ICCMC)*, pp. 442-447.

[8] Colak, I., and Yildirim, D., (2009) Evolving a Line Following Robot to use in shopping centers for entertainment, Industrial Electronics, *35th Annual Conference of IEEE*, pp.3803 – 3807.

Chapter 3

A study of parametric investigation for determining the effect of sustainable cryogenic treated ceramic insert to machine Inconel 718

Ivan Sunit Rout, P Pal Pandian, Anil Raj, Ramesha K, Anil Melwyn Rego and Darshan S M*
Department of Mechanical and Automobile Engineering, School of Engineering and Technology,
CHRIST (Deemed to be University), India

Abstract

Inconel 718 is a nickel-chromium based super alloy and is one of the hardest material to cut. It has high thermal resistance, high creep, highcorrosion resistance and retains toughness and strength at elevated temperatures. It has an excellent yield strength even at elevated temperatures and it is due to these chemical and mechanical properties the tool life is extremely short. This hard to cut metal has a wide scope in the field of bio medical industry, aerospace industry, bearing industry, steam turbine and nuclear applications andthe demand has rapidly increased in the recent years. Ceramic tool is one such cutting tool being used in the machining of this metal and study is still being conducted to increase the machinability. The objective of this study is to investigate the parameters of cutting to determine themachining characteristics of Inconel 718 using cryogenic treated ceramic tool based on Grey Relation Analysis (GRA) and signal to noise (S/N) ratio. The input parameters such as feed rate, cutting speed and depth of cut are taken into account to obtain the optimum output parameters such as minimal surface roughness and low tool wear rate to improvise the machining characteristics of this superalloy.

1. INTRODUCTION

Inconel718 is a high-strength, corrosion-resistant nickel basedsuper alloy suited for applications requiring high temperatures with favorable characteristic features. At very high temperatures, it has high oxidation and corrosion resistance properties. The alloy also retains a high mechanical strength under high temperature conditions. Ithas a wide scope in the field of bio medical industries, aerospace, bearing industry, steam turbine and nuclear applications and the demand has rapidly increased in the recent years. However, Inconel718 is classified based on its difficulty to cut materials due to its physical properties and rapid work hardening. Tool failure and rapid tool wear in machining has been identified as a challenging problem.

The thermal conductivity of Inconel 718 compared to that of a commonly used alloy steels is much lower. Hence, the cutting temperature in the tool and the work piece are notably high during machining of these materials.poor choice of machining parameters causes cutting tools to wear and break swiftly. The damaged work piece and poor surface quality is not cost effective. Therefore, to intensify the machining performance in cutting these materials, the machining parameters should be optimized.

The main objective of this paper is to investigate the cutting parameters of Inconel 718 super alloy using cryogenically treated ceramic inserts to reduce machining difficulties. The ceramic insertsare cryogenically treated to improve wear resistance and reduce residual stresses. In addition to seeking stabilization and enhanced relief, or resistance of tool wear, the corrosion resistance also gets improved.The study explores the parameter optimization for Inconel 718 nickel-based super alloy with the Grey Relation Analysis and analysis of variance (ANOVA) for signal to noise ratio (S/N). Considering various characteristic features, the three optimized parameters are: feed rate, cutting speed, and depth of cut. Here with the use of cryogenically treated ceramic inserts the machining performance can be improved effectively for the machining of Inconel 718 in order to obtain minimal surface roughness and low tool wear.

2. EXPERIMENTAL SET-UP

2.1 Methodology

The experiment is conducted on a 3-axis CNC vertical milling machine under wet machining conditions. The machine is equipped with a motor of 25 kw whose feed rate is 10m/min and spindle speed is 1800 rpm.

24

Fig 1 Setup of CNC vertical milling machine

A fixture is specifically prepared to hold the workpiece in order to reduce the noise and vibrating conditions. Thefixture consists of a main plate, support plate and a slot plate. All the holes in the fixtureare M12 tapped hole.

The dimensions of the fixture are:
• Main plate : 200x200x12mm
• Support plate: 80x12x50mm
• Slot plate : 100x100x12mm
• Slot size : 100x80x3mm
• Hole : M12 tapped holes

Fig 2 Fixture of the workpiece (Inconel 718)

The workpiece used is Inconel 718 in the form of a block whose dimension is 75mm length, 60 mm breadth and 50 mm height. The insert used is ceramic of AS20 grade which is round in size and described by RPGX 1204 CH AS20.

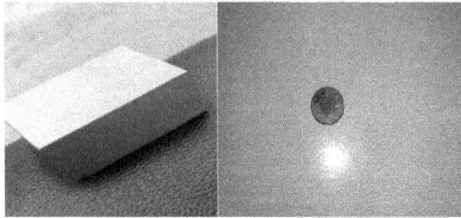

Fig 3 Workpiece (Inconel 718)   Fig 4 Cutting tool insert of ceramic

2.2 Equipment

Cryogenic Treatment System

The stainless steel container which is a multi-walled has a cryotreatment unitwhose inner spaceis made up of Polyurethane foam. The upper cover has a double end shaft which is stainless stell and the fan-motor assembly is centrally mounted. The liquid nitrogen connects the feed through along with valve,outlet connections and pressure gauges are all mounted on the upper cover. Under the upper cover, a buffer tank of cylindrical shape is mounted where the liquid nitrogen gets accumulated which is supplied continuously and thevapours are vented through a vent pipe which gets vaporized. The fan motor assembly shaft passes through the buffer tank and on both the sides, mounted are the fan blades. To ensure better transfer of heat, at the bottom of the buffer tank is fixed copper disc. Aluminium fins are fixedcircumferentially to the bottom of the copper disc to ensure that forced convection inside the chambergets proper cooling in space where cryotreatment has undergone.

When metals are cooled to cryogenic temperatures gradually and soaked for a longer period and then warmed to ambient temperature at a predetermined rate, whereby the structure of the lattice atoms changes due to stresses being relieved during treatment of cryogenic cycle. A typical cryotreatment cycle involves involves mainly three phases where the metal samples are cooled gradually

26

to cryogenic temperature (cooling period), holding for prolonged length of time (soak time) and gradually warming (warming time) to ambient temperature.

Fig 5 Cryogenic treatment Unit

Tool Maker's Microscope

This is an instrument which is used to measure the tool wear in millimeter. The glass plate (stage glass)on which the specimen is placed where the source of lightis fixed underneath, and connected to the guiding screw of the plate with handlefor the adjustments.the image view from the scope has thin line of cross and thecenter of the cross is to be set on the starting point of the measurement. The end line of the grey surface shown on themicroscopefor thetool wear measurementis set as the starting point and the distance to the end of the circumference of thetool insert is measured by rotating vertical handle.

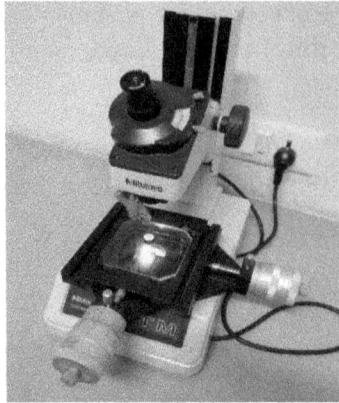

Fig 6 Mitutoyo Tool Maker's Microscope

Surface Profilometer

Surface profilometer is the measuring instrument which is used to measure the roughnessof a surface and gives the result as the average distance between the two peaksof the regular profile. The equipment consists of measuring tip, cantilever transducerand a control system to plot the profile of surface in 2D graph and evaluate the mean value of it. It gives the value in microns.

Fig 7 Measurement of roughness by surface profilometer

2.3 Design of experiments

The experiment is carried out based on a taguchi L9 orthogonal array taking cutting parameters as spindle speed, feed rate and depth of cut while response parameters are surface roughness and tool wear.

Table 1 Cutting parameters and their levels

Parameter	Unit	Symbol	Level 1	Level 2	Level 3
Cutting Speed	rpm	v	500	600	700
Feed Rate	mm/min	f	10	15	20
Depth of cut	mm	d	0.2	0.4	0.6

Then analysis was carried out on table 2 using grey relational analysis (GRA) and analysis of variance for S/N ratios.

Table 2 Cutting and response parameters based on L9 array

Runs	Cutting speed (rpm)	Feed rate (mm/min)	Depth of cut (mm)	Surface Roughness (µm)	Tool wear (mm)
1	500	10	0.2	1.01	0.085
2	500	15	0.4	1.23	0.102
3	500	20	0.6	1.19	0.042
4	600	10	0.4	0.81	0.022
5	600	15	0.6	1.53	0.030
6	600	20	0.2	1.39	0.365
7	700	10	0.6	1.02	2.180
8	700	15	0.2	1.24	1.220
9	700	20	0.4	1.17	0.877

3. RESULTS and DISCUSSION

3.1 GRA Analysis

Setting up the Eigen value matrix and input original data

$X_1 (1) \quad X_1 (2) \quad X_1 (3)....... X_1 (n)$
$X_2 (1) \quad X_2 (2) \quad X_2 (3)....... X_2 (n)$

$$X = \begin{pmatrix} X_3 (1) & X_3(2) & X_3 (3)....... X_1 (n) \\ : & : & : & : \\ : & : & : & : \end{pmatrix} \qquad (1)$$

$X_m (1) \quad X_m(2) \quad X_m (3)....... X_m (n)$

where m −no. of listed experimental runs n −no. of influence factors $X_m(n)$ −corresponding value of the m experimental run and n influence factor

Based on the equation (1) the entire data is formulated as table 2 in which control factors are cutting speed (A), feed rate (B) and depth of cut (C) and response variables are surface roughness (X) and tool wear (Y).

Transformation of observed data into standardized data using formulas

After the first step all the observed data are subjected to a transformation process which is based on the influence factors. The influence factors are surface roughness and tool wear. The defect type factor is used which states the smaller the better.

$$\text{Defect type factor } Xm(n) = \frac{\max|Xm(n)|-Xm(n)}{\max|Xm(n)|-\min|Xm(n)|} \qquad (2)$$

Table 3 Transformation of observed data into standardized data using equation (2)

	Observed Data		Standardized Transformed Data	
Experimental Run	Surface Roughness (μm)	Tool wear (mm)	Surface Roughness (μm)	Tool wear (mm)
1	1.01	0.085	0.72	0.971
2	1.23	0.102	0.42	0.963
3	1.19	0.042	0.47	0.991

4	0.81	0.022	1.00	1.000
5	1.53	0.030	0.00	0.996
6	1.39	0.365	0.19	0.841
7	1.02	2.180	0.71	0.000
8	1.24	1.220	0.40	0.445
9	1.17	0.877	0.50	0.604

Calculation of grey relational degree and ranking

Selection of the best value for each influence factorThe best value for each influence factor is selected. Surface roughness and tool wear should be less. Hence the least value of surface roughness and tool wear is selected and chosen as the referential series for further calculation.

Table 4 Referential series from standardized data

Experimental Run	Surface Roughness (μm)	Tool wear (mm)
X_0	0.19	0.445
1	0.72	0.971
2	0.42	0.963
3	0.47	0.991
4	1.00	1.000
5	0.00	0.996
6	0.19	0.841
7	0.71	0.000
8	0.40	0.445
9	0.50	0.604

Calculation of Grey Relational Degree

- Selection of the best value for each influence factor.
- Getting the absolute difference of compared series and referential series.
- Find out the minimum and maximum.
- Choose the constant p = 0.4 (surface roughness) and 0.6 (tool wear).

- Calculation of relational co-efficient and relational degree using equation (3) and (4)

Relation coefficient $\tau_m(n) = \dfrac{\Delta min + p\Delta max}{\Delta X_m(n) + p\Delta max}$ (3)

Relational degree $r_m(n) = \Sigma| \omega\tau_m|$ (4)

Table 5 Relational degree and the ranking

Experimental Run	Surface Roughness (μm)	Tool wear (mm)	Relational Degree	Rank
ω	0.4	0.6	-	-
τ_1	0.60	0.573	0.584	8
τ_2	0.93	0.578	0.719	5
τ_3	0.85	0.559	0.675	6
τ_4	0.45	0.554	0.512	9
τ_5	1.00	0.556	0.734	4
τ_6	1.59	0.675	1.041	2
τ_7	0.61	0.632	0.623	7
τ_8	0.96	1.477	1.270	1
τ_9	0.81	1.000	0.924	3

3.2 ANOVA Analysis for S/N ratios

It was conducted in the minitab by analyzing the table 4 for determining S/N ratios taking the smaller the better factor into consideration for surface roughness and tool wear.

Table 6 Analysis of variance for S/N ratios of surface roughness

Source	DF	Seq SS	Adj SS	Adj MS	F	P
Cutting speed (rpm)	2	0.3896	0.3896	0.1948	0.15	0.872
Feed rate (mm/min)	2	15.0907	15.0907	7.5454	5.71	0.149
Depth of cut (mm)	2	3.2017	3.2017	1.6008	1.21	0.452

Residual Error	2	2.6432	2.6432	1.3216		
Total	8	21.3252				

Note: S=1.1496   R-sq= 87.71%   R-sq (adj)= 50.42%

The analysis from table 6 shows that all the factors have significant contribution for surface roughnes as all P-Value > 0.05 accept-ingalternate hypothesis. However Feed rate has a major and most significant contribution in determing the surface roughness whose F-value is 5.71 and can be also known from response table 7. However, the response variable surface roughness predicts better effect than tool wear on the machining of Inconel 718 (R-sq (adj) = 50.42%).

Table 7 Response Table for S/N ratios of surface roughness

Level	Cutting speed (rpm)	Feed rate (mm/min)	Depth of cut (mm)
1	1.1318	0.5240	1.6051
2	1.5746	2.4535	0.4438
3	1.1347	1.9117	1.7923
Delta	0.4428	2.9774	1.3484
Rank	3	1	2

Table 8 Analysis of variance for S/N ratios of tool wear

Source	DF	Seq SS	Adj SS	Adj MS	F	P
Cutting speed (rpm)	2	1351.56	1351.56	675.78	5.09	0.164
Feed rate (mm/min)	2	25.76	25.76	12.88	0.10	0.912
Depth of cut (mm)	2	131.89	131.89	65.95	0.50	0.668
Residual Error	2	265.38	265.38	132.69		
Total	8	1774.60				

Note: S=11.5192   R-sq= 85.05%   R-sq (adj)= 40.18%

33

The analysis from table 8 shows that all the factors have significant contribution for tool wear as all P-Value > 0.05. However cutting speed has a major and most significant contribution in determing the tool wear whose F-value is 5.09 and also can be noticed in response table 9. However, the response variable tool wear predicts smaller effect on the machining of Inconel 718 (R-sq (adj) = 40.18%).

Table 9 Response Table for S/N ratios of tool wear

Level	Cutting speed (rpm)	Feed rate (mm/min)	Depth of cut (mm)
1	22.925	15.931	9.480
2	24.121	16.186	18.040
3	-2.452	12.476	17.074
Delta	26.573	3.710	8.560
Rank	1	3	2

3.3 Main Effect Plots for machining parameters

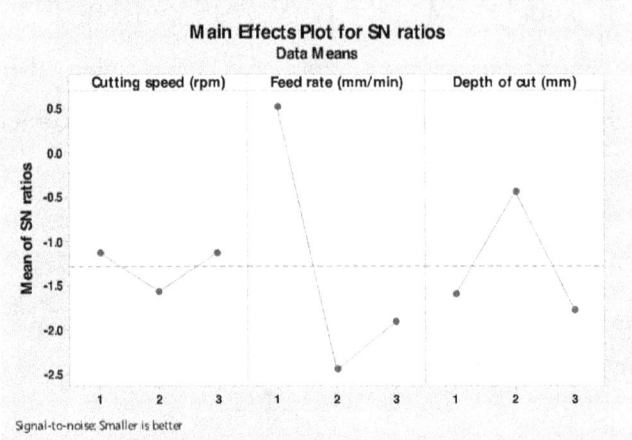

Fig 8 Main Effect plot of S/N ratio for surface roughness

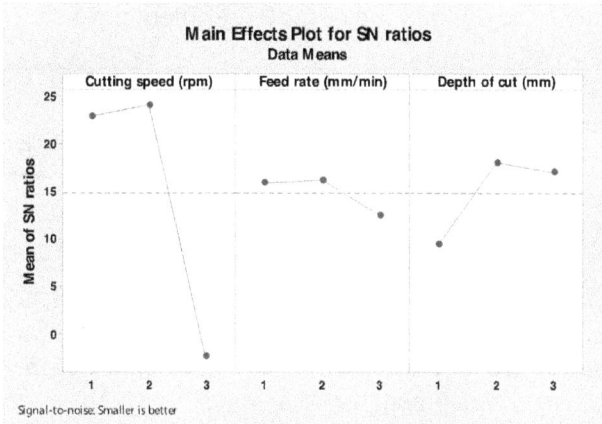

Fig 9 Main Effect plot of S/N ratio for tool wear

In figure 8, it can be seen that cutting speed 700 rpm, feed rate 20 mm/min and depth of cut 0. 2 mm has better optimal means than the rest as they are closest to the mean lines for surface roughness while in figure 9, it can be seen that cutting speed 500 rpm, feed rate 10 mm/min and depth of cut 0.6 mm has better optimal means than the rest as they are closest to the mean lines for tool wear.

4. CONCLUSION

In improving the machinability of Inconel 718, cryogenic treated ceramic insert plays a major role in enhancing the cutting parameters and providing minimal surface roughness and low tool wear. It can be seen by GRA analysis that cutting speed 700 rpm, feed rate 15 mm/min and depth of cut 0.2 mm prove to be better than the rest experimental runs from table 10.

Table 10 Ranking of the experimental runs

Rank	1	2	3	4	5	6	7	8	9
Experimental Run	8	6	9	5	2	3	7	1	4

ACKNOWLEDGEMENT

The authors are grateful and honoured to CHRIST (Deemed to be University) for providing all types of financial means and aids in completing this work. This work has been carried out in Advanced Machining Lab, Department of Mechanical and Automobile Engineering, Faculty of Engineering. The authors appreciate the lab facilities and support provided by the CHRIST fraternity in making this work a successful one.

REFERENCES

[1] Shokrani A., Dhokia V. and Newman S. T. (2017) Hybrid cooling and lubricating technology for CNC milling of Inconel 718 nickel alloy,*27th International Conference on Flexible Automation and Intelligent Manufacturing*, **11**, pp. 625-632.
[2] Dumasi M. C. A. and Kulkarni D. V. A. (2017) Effect of cryogenic treatment on en8 steel used for press tool, *Global Journal of Engineering Science and Research Management*, pp. 39-54.
[3] Chopra S. A. and Sargade V. G.(2015) Metallurgy behind the Cryogenic Treatment of Cutting Tools: An Overview, *4th International Conference on Materials Processing and Characterization*,*2*, pp. 1814-1824.
[4] Rout I.S. and Pandian P.P. (2018) Selection of cutting parameters for the machinability of Inconel 718 using Grey Relational Analysis, *MATEC Web of Conferences*, **172** (04004).
[5] Schornika V., Zetekb M., Danac M. (2015) The Influence of Working Environment and Cutting Conditions on Milling Nickel – Based Super Alloys with Carbide Tools, 25th *DAAAM International Symposium on Intelligent Manufacturing and Automation, Procedia* Engineering 100, pp 1262 – 1269.
[6] Xavior, Manohar, Jeyapandiarajan and Madhukar P.M. (2017) Tool wear assessment during machining of inconel 718, *13th Global Congress on Manufacturing and Management*,*174*, pp 1000-1008.

[7] Ramya R., Rout I.S., Yongmin L., Kumar M. and John A.J. (2018) Optimization of Cutting Parameters for the Machining of Inconel 718 using Grey Relational Analysis, *Journal of polymer and composites, STM Journal*, **6** (2), pp 1-5.

[8] Kaynak Y. (2014) Evaluation of machining performance in cryogenic machining of Inconel 718 and comparison with dry and MQL machining, *International Journal of Advanced Manufacturing Technology*, **72**, pp 919-933.

[9] Akhtar W., Sun J., Sun P., Chen W. and Saleem Z. (2014) Tool wear mechanisms in the machining of Nickel based super-alloys: *A review, Frontier Mechanical Engineering*, **9**(2), pp 106-119.

[10] Zhuang K., Zhang X., Zhu D. and Ding H. (2015) Employing preheating- and cooling-assisted technologies in machining of Inconel 718 with ceramic cutting tools: towards reducing tool wear and improving surface integrity, *International Journal of Advanced Manufacturing Technology*, **80**, pp 815-1822.

Chapter 4

MPPT METHOD FOR PV SYSTEMS UNDER PARTIALLY SHADED CONDITIONS BY APPROXIMATING I-V CURVE

S. Selvakumaran and K. Baskaran
EEE, Alagappa Chettiar Government College of Engineering and Technology, Karaikudi, India

ABSTRACT

The environmentally friendly energy based on solar energy explains the fascinating choice response of traditional resources on the earth. In addition, as the cost per watt of solar cell modules decreases, solar cell modules are becoming more and more attractive today. The yield voltage provided by the solar cell module is very low. In this way, the series connection of various modules is used for proper continuous voltage performance. To a certain extent hidden photovoltaic clusters have different photovoltaic brand peaks. The technology described in this article implements GMPP in a deterministic and exceptionally fast way. Use the cluster's P-V curve to cite some examples wisely, and divide the hunting voltage range into smaller sub-regions. Then, at this point, the I-V curve of each sub-region is approximated by the base curve, and the additional limit of the cluster power in the sub-region is similarly measured. Therefore, by comparing the intentional true force estimate with the limits of further evaluation, the GMPP hunting ground is limited, and the sculpted area is limited according to some characteristic rules. The correlation between entertainment and search results is introduced to characterize the expression and superiority of the proposed method.

1. INTRODUCTION

Solar energy is free, unlimited and clean. Very attractive stock alternatives for modern and regional applications have incredible potential, especially in remote areas, such as water siphons, heating and cooling [1]. Photovoltaic structures use solar cell modules to convert sunlight into electricity. Due to the advantages of fewer prerequisites for support, lower fuel costs, and no chaos due to lack of moving parts, the solar age has gained broader significance as a sustainable energy source. Recently, photovoltaic siphon structures have received more research, especially in remote areas where matrix coupling is practically impractical or expensive. In addition, due to recent major advances in the field of materials and solar cell innovation, photovoltaic siphons have also received extensive consideration [2-3].

They are widely used in livestock and livestock water storage and limited water system structures. After controlling these solar-based solar structures, the most extreme intensity is worth the manufacturing activity. The measured value of solar radiation captured from the surface differs in the cosine of the origin between the sun's rays and the general rays of the surface. The plot points of the plane are called vertices. The sun shines in better places, in multiple places. By expanding the range, the curvature of the earth reduces the sun apex observed in the sky. The cluster must also be inclined to compensate for this impact [4-5].

The PV cluster moving south from a point in the φ range has a solar θ generation point that is similar to the PV exposure level in the θ-β range. In other seasons, due to the mysterious mathematics applicable to the earth's solar conditions, the character is expected to follow the sun for solar energy exhibitions and collections. You can implement a global positioning framework for horizontal or thinking solar modules and heat engines. However, structures that rely on concentrated sunlight can generally only recognize light that passes through narrow points. They usually waste scattered parts based on the sun to use the direct part of the sun. This balances the benefits of these global positioning structures, which usually continuously capture the thickness of the most common extreme forces in sunlight [6-7].

A dimensionless coefficient, called the coefficient of variation, is used to calculate the solar radiation on the displacement surface of the photovoltaic cluster in any area from the solar radiation on the plane. The displacement coefficient is the percentage of cosine at the cosine frequency point of the vertex or the percentage of solar radiation on the plane displacement surface relative to the season, distance, and tilt point. Displacement, similar to the century of solar radiant energy, varies by region, season, and slope. The most commonly used climate information when planning a solar structure is sunshine, ambient temperature and wind speed information. The structure of the sun is very sensitive to climate change, and the amount of exposure should be changed appropriately. In addition, since long-term climate conditions cannot be assessed, it is expected that historical climate data will be carefully managed to plan solar panels [8-9].

2. RELATED WORKS

In [10] Pedro Neves, D. Gonçalves, J. G. Pinto, Renato Alves, Joao L. Afonso et al presents A single shunt active filter step associated with a set of solar panel and combined with maximum power point tracking. The intensity level of the active shunt filter consists of a dual-arm IGBT inverter driven by a digital signal processor, which can be controlled according to the instantaneous reactive power theory (p-q theory). MPPT depends on the distribution circuit indicated by the DSP running the MPPT algorithm. The output of the MPPT circuit is connected to the DC side of the parallel active filter. The switchboard can pay for the power factor and the sound of the flow, and at the same time use a similar inverter to inject the electric energy provided by the MPPT-managed solar direction panel into the force matrix. Introduce the side effects of equipment operating in power plants under various conditions, as well as the configuration and details of the equipment.

In [11] Jubaer Ahmed, Zainal Salam et al presents For photovoltaic panels, the slightly hidden vertex of the photovoltaic curve is usually expected to be the product of 0.8 Voc, where Voc is the open circuit voltage of the photovoltaic module. However, this assumption called the semi-hidden 0.8 Voc model is not clear. If a similar model is used to plan all the maximum intensity point followers, the calculation will illuminate the unauthorized area of the PV curve to find the wrong position at the top of the world. This article tries to show the deficiencies of the 0.8Voc model as the number of string modules increases. The work also recommends direct reports to predict the correctness of Pinnacle. For 20 module (400V Voc) lines, the maximum deviation between the actual vertex and the calculated vertex is less than 3V, while the deviation of the 0.8Voc model is as high as 50V. In addition, according to the proposed strategy, the effect of MPPT can work about 2%.

In [12] Yunping Wang, Ying Li, Xinbo Ruan et al presents In a slightly hidden state, PV strings represent complex performance attributes. In other words, the current-voltage curve shows each step of the current, and the force-voltage curve represents multiple maximum forces. Therefore, the general strategy of tracking the most extreme Himjom is inappropriate in terms of accuracy and tracking speed. This article provides the skip-judge global MPPT survey of the global MPPT survey and the Quick Worldwide MPPT technology on the step-down side. This relies on two global MPPT strategies, especially the in-depth investigation of PV string IV and PV characteristics. SSJ-GMPPT technology can track the true ultimate strength even under certain hidden conditions, and can achieve high accuracy and optimized speed without additional circuits and sensors.

In [13] Yihua Hu, Wenping Cao, Jiande Wu, Bing Ji and Derrick Holliday et al presents Another guide for more extreme forces, based on thermal imaging measurements, follows a plan to solve the problem of photovoltaic incomplete concealment. In the age of solar power generation, it is like a serial drama, which uses a large number of photovoltaic cells connected to a cluster, which actually circulates in a huge field. At the moment when the photovoltaic module is blamed or partially hidden, the photovoltaic structure sees the irregular circulation of power generation and thermal contours, and various fascinating times are concentrating. If left untreated, extreme problems may occur, reducing overall strength and damaging the structure. In this article, use a thermal camera at the defect location and generate other MPPT diagrams to change the operating point to adjust the advanced MPP. Extract rich information from thermal camera images to find MPPs around the world. This can be used to reduce the MPPT time and check the MPP reference voltage.

In [14] Kai Chen, Shulin Tian, Yuhua Cheng and Libing Bai et al present The subsequent maximum intensity is an integral part of the energy converter that utilizes the photovoltaic display. There are some major advantages near the normal force voltage of photovoltaic clusters operating in a semi-hidden state. In this article, the world's most powerful (PV. Contrasted and World MPPT Procedures) previously proposed by World MPPT Procedures determines whether to half-hide the strategy proposed in this article, with the peak of the solar curve. GMPP and LMPP area prediction new strategies can quickly discover GMPP and avoid many energy misfortunes caused by blind scanning.

3. PROBLEM DEFINITION

The most commonly used climate information for planning solar structures are copies based on solar energy, temperature and wind speed information. Solar panels are very sensitive to climate change, and their expressions will also change. Similarly, long-term climate conditions cannot be inferred, so it is expected that there will be intentional management of historical climate data to plan solar panels. This information should include all reasonably predictable long-term figures. Less complex strategies may use general remote climate information. However, the usual message is that in facilities that do not reflect sunny days, the expected daily prerequisites must be met in a stable manner. These strategies have experienced the negative effects of dynamic complexity and large changes in the execution frame, and require different sampling and lower MPPT speed at different points of the P-V mark.

4. PROPOSED SYSTEM

In this interaction, he proposed another calculation to find GMPPT under all irradiation conditions. The proposed strategy divides the voltage detection area into smaller sub-earths, taking the I-V curve as an example, and measures the maximum power cut-off of the cluster locally. Then, compare the most significant inspection power at the time with the maximum confirmation limit to narrow the right to pursue. No zone is set through this system. Finally, the specific GMPP is followed by P&O calculations.

4.1 BLOCK DIAGRAM

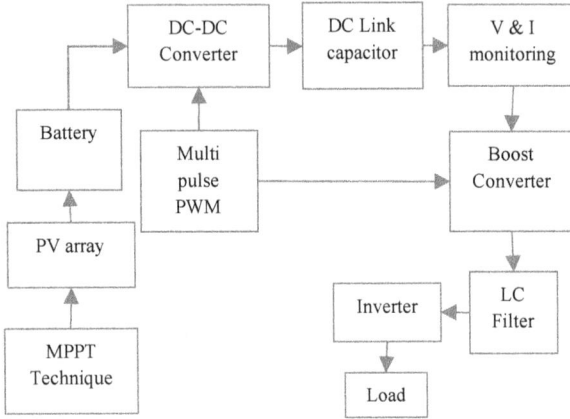

Fig 1 Block diagram

4.2 PROPOSED PROCESS EXPLANATION

4.2.1 SOLAR PANEL

The solar device board is a square data measurement format, which is light in weight and suitable for power consumption. They are often referred to as "photovoltaic" panels because the largest solar source available is the sun, and free consultants refer to them as brushes. All other researchers use photovoltaic cells to describe it, and photovoltaic cells use "light energy" as the most important capital. Sunny planks may be a social accident of solar cells. Around many houses in the outer regions of the world, solar cells help generate enough energy. The additional energy generated by the auxiliary light to call the mobile phone is a sun, which is continuously pointed at the sun through the general movement of the rocket body, so that the chest wall of the tank can be drilled out of the tank without being connected. The shifted overtime time can be calculated on the panel. There are tanks everywhere.

Fig 2 Solar panel

4.2.2 MAXIMUM POWER POINT TRACKING SYSTEM

The next most extreme intensity is the process of using grid-related inverters, solar chargers, and similar devices to obtain the maximum intensity imaginable in one or more solar modules. Photovoltaic solar cells have an embarrassing connection between temperature and the absolute interference of solar energy research. These studies provide non-linear yields that can be divided along the IV curve. The basic principle of the MPPT framework is to test the output of solar cells and to apply legal protection to the most extreme power gains under certain arbitrary ecological conditions. The MPPT unit is usually tuned to the converter frame to provide voltage or current changes, main body, and guidance to drive various loads, including grids, batteries, or motors.

Fig 3 MPPT process

4.2.3 I-V, P-V CURVE AND MAXIMUM POWER POINT

The I-V and PV curves of the 30W PV module simulate the MATLAB model. Photovoltaic modules can generate force in terms of operating points at all points on the I-V curve. The direction of the operating point is the operating voltage and operating current. There is an interesting point near the inflection point of the I-V curve, called MPP (Maximum Force Point). At this point, the module runs at its maximum capacity and produces the most extreme yield force. As shown in the figure, the force-voltage diagram is nested in the PV module I-V diagram. He found that the measured value of the force produced by the solar cell module changes surprisingly depending on the working conditions. Use the MPP processing structure of the photovoltaic module to take advantage of the extreme strength of the module.

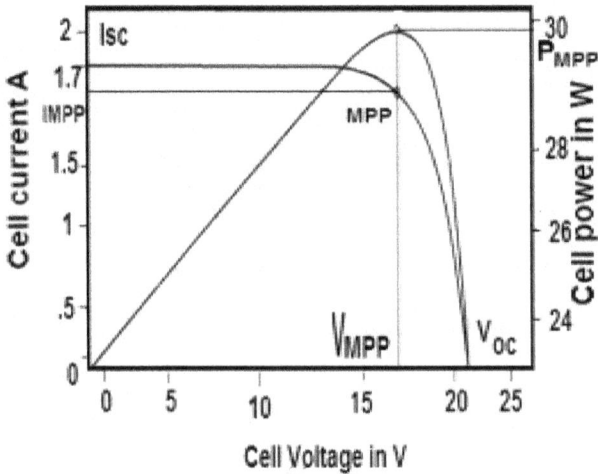

Fig 4 I-V Curve

4.2.4 CONTROLLER UNIT

Microcontrollers are small microcontrollers designed to drive structures contained in high-speed vehicles, robots, related machines, multi-component medical equipment, multi-function cordless phones, candy machines, household appliances, and various other devices. It is a computer. Registered trademarked microcontrollers include processors, memories, and peripherals. The simplest micro-controller can easily create the activities of electromechanical pan-els mounted on common practical panels. In the early days, this use was restricted by large machines such as car heaters and engines to optimize productivity and execution.

4.2.5 BATTERY

A battery is a group of devices used to supervise single or additional synthetic scientific cells, which have subtle and insignificant rela-tionships that affect the development of electricity, such as spot-lights, large phones, and electric cars. When the battery is powered, the positive terminal is the negative terminal, and the unnecessary terminal is the positive terminal. A physical obstacle that can be de-tected is that an electronic store that can access an external circuit once can transfer energy to non-essential equipment.

Every time the battery is connected to the input external circuit, the square electrolyte prepares to move to the internal particles, termi-nates the weft response from other terminals, and transfers energy to the external path. Agreeing to operate the battery is a human de-velopment of these particles in the battery. Generally speaking, the term "battery" is used for quick decision-making machines to adjust the lush cells; however, this program has a different review from the program that accepts cell settings gadgets.

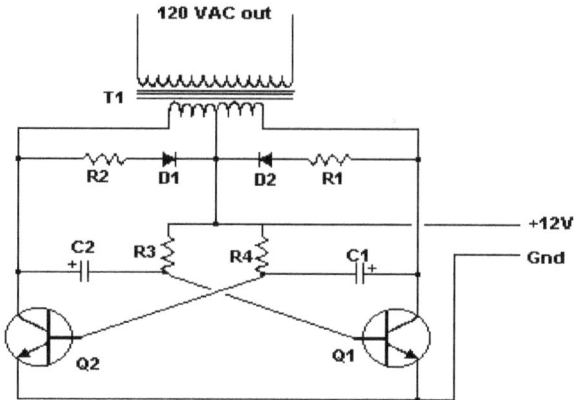

Fig 5 Battery

47

4.2.6 DC-DC CONVERTER

DC-DC converters are used in portable electronic devices, such as batteries and PCs, and are mostly driven by batteries. These electronic devices usually include several sub-circuits, each of which is provided by a battery or an external power source (sometimes higher or lower than the stock voltage). It has its own prerequisites, and the voltage levels are not exactly the same. In addition, the battery voltage will decrease as the stored energy is exhausted. The improved DC-DC converter provides a loop and in some way boosts the battery voltage that is not sufficiently reduced, which is indistinguishable compared to the use of various serial connections and saves space. Most DC-DC converters monitor the build voltage.

It is a mix of high-efficiency LED power supplies, which are various DC-DC converters that manage the current through the LED, and a prudent protection trap that can double or triple the breakdown voltage. Urbanized DC-DC converters used to harvest electricity from photovoltaic panels and wind turbines are called authorized devices. The modification used to trade the voltage on the main frequency of 50-60Hz should be huge and heavy, with a few watts or more of power. Due to eddy currents from the center, this tends to raise them and cause pain to the windings. DC-DC systems that use transformers or inductors operate at frequencies that other systems cannot match, so only more moderate, lighter, and cheaper winding components are required.

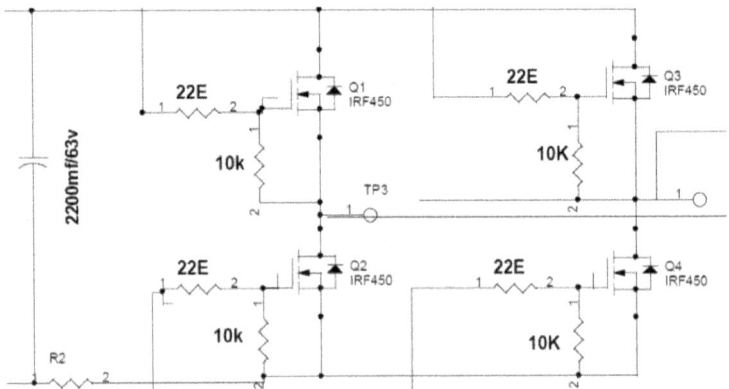

Fig 6 DC to DC Converter circuit

48

4.2.7 Boost Converter

The lift converter is a DC-DC power converter, and its breakdown voltage is more obvious than the beneficial voltage. It is a switch-mode power supply that contains two semiconductors in each case, and in each case, the energy-saving components are capacitors, inductors, or a mixture of the two. A channel composed of capacitors is usually added to the output of the converter to reduce the increase in the output voltage.

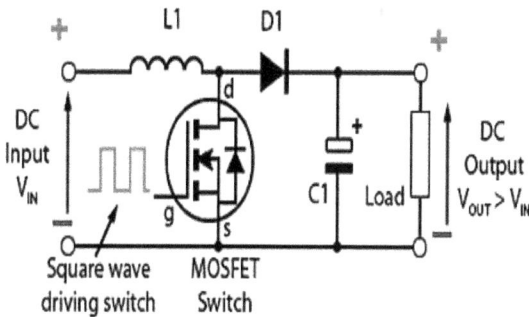

Fig 7 Boost converter

The key rule that drives the lift converter is the inclination of an inductor to oppose changes in current by making and obliterating an attractive field. In a lift converter, the yield voltage is consistently higher than the info voltage.

(a) When the switch is shut, electrons course through the inductor clockwise way and the inductor stores some energy by producing an attractive field. Extremity of the left half of the inductor is positive.

(b) When the switch is opened, current will be decreased as the impedance is higher. The attractive field recently made will be annihilated to keep up the current towards the heap. Accordingly the extremity will be switched. Thus two sources will be in arrangement making a higher voltage charge the capacitor through the diode D.

4.2.8 INVERTER

It is an electronic device or hardware that changes direct current (DC) to alternating current (AC). The information voltage, yield voltage and recurrence, and by and large force dealing with rely upon

49

the plan of the particular gadget or hardware. The inverter doesn't deliver any force; the force is given by the DC source. A force inverter can be totally electronic or might be a blend of mechanical impacts (like a rotational contraption) and electronic hardware. Static inverters don't utilize moving parts in the change cycle.

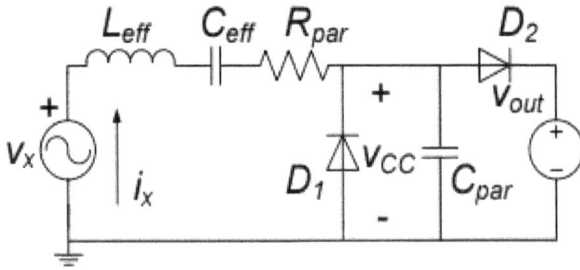

Fig 8 Inverter circuit

4.2.9 DC LINK CAPACITOR

A DC interface is an association which interfaces a rectifier and an inverter. These connections are found in converter circuits and in VFD circuits. The AC supply of a particular recurrence is changed over into DC. This DC, thus, is changed over into AC voltage. The DC interface is the association between these two circuits. The DC connects for the most part has a capacitor known as the DC interface Capacitor. This capacitor is associated in equal between the positive and the negative conductors. The DC capacitor keeps the homeless people from the heap side from returning to the merchant side. It additionally serves to smoothen the beats in the amended DC.

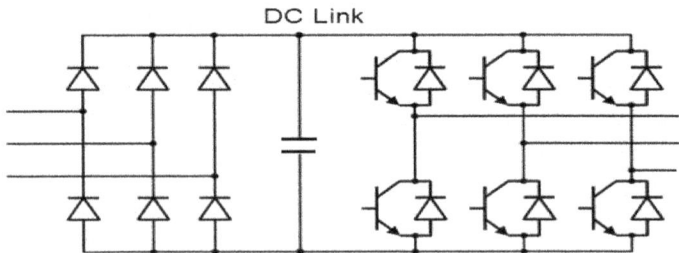

Fig 9 DC Link Capacitor

4.2.10 LC FILTER

An LC circuit, also called a resonant circuit, a storage circuit, or a tuning circuit, is a circuit composed of an inductor denoted by the letter L and a capacitor denoted by the letter C. In addition to the energy that swings in large iterations of the circuit, the circuit can also act as an electric resonator for a simple electric tuning fork. The LC circuit is used to generate a signal from a specific iteration and select symbols with more complex symbols in the specific iteration. This function is called bandpass channel. They are an important part of many electronic devices, especially wireless hardware, used in circuits such as oscillators, channels, tuners, and repeat mixers.

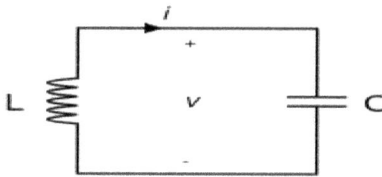

Fig 10 LC Filter

4.2.11 PWM PROCESS

Pulse width modulation (PWM) is a powerful program for controlling simple circuits with high-volume processors. This is a very useful process, widely used in the field of correspondence between power control and conversion estimation. This program is used to reduce the THD (Total Harmonic Distortion) of the load current. Total harmonic distortion or THD is especially important for the strength of the keys, just the strength of all chorus parts. In this strategy of checking inverter switches with sinusoidal PWM output, regular sinusoidal reference signals (balance or control signals) and trilateral fine transmitter signals are attractive for drawing repeated exchanges.

Multiple Pulse Width Modulation (MPWM)

The significant impediment of single PWM practice is high consonant substance. To reduce the consonant substance, the various PWM procedures is utilized, in which a few heartbeats are given in each half pattern of yield voltage. The symphonious fulfilled can be dense by utilizing successive heartbeats in every half pattern of creation voltage the gating signals are designed by assess reference signal with three-sided transporter wave

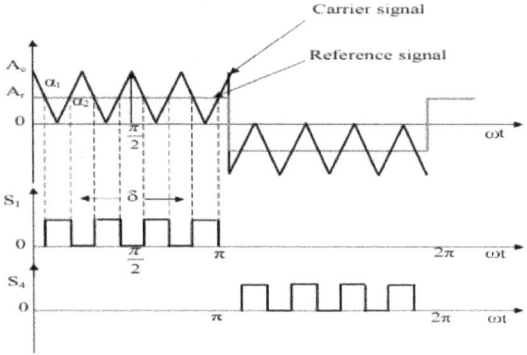

Fig 11 Pulse width modulation

4.2.12 LCD Display unit

LCD is used to visualize a collection of devices in an enterprise. We used a 16-character alphanumeric display case, showing 16 sections and 2 columns. Therefore, 16 characters are printed per line. Therefore, enter 32 characters to display in the 16x2 alphanumeric demo. The alphanumeric display cabinet can also be used to demonstrate the performance of various modules connected to the microcontroller. Therefore, the alphanumeric display cabinet plays an important role in the process of fixing the square modules, it only sends out notifications when the frame starts to disappoint, because it depicts surrender and identifies obstacles.

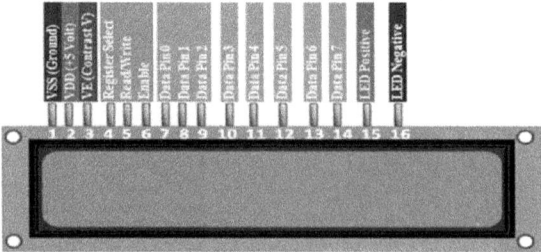

Fig 12 LCD display unit

5. RESULTS AND DISCUSSION

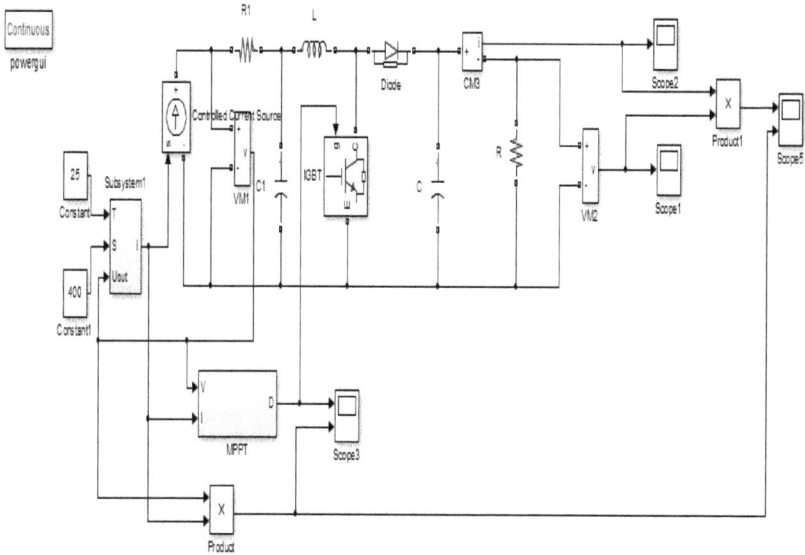

PV Array-MPPT circuit

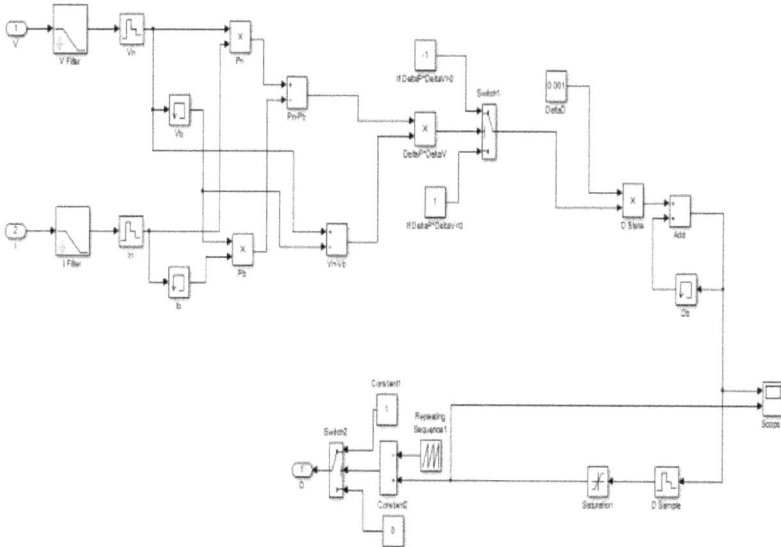

MPPT Output

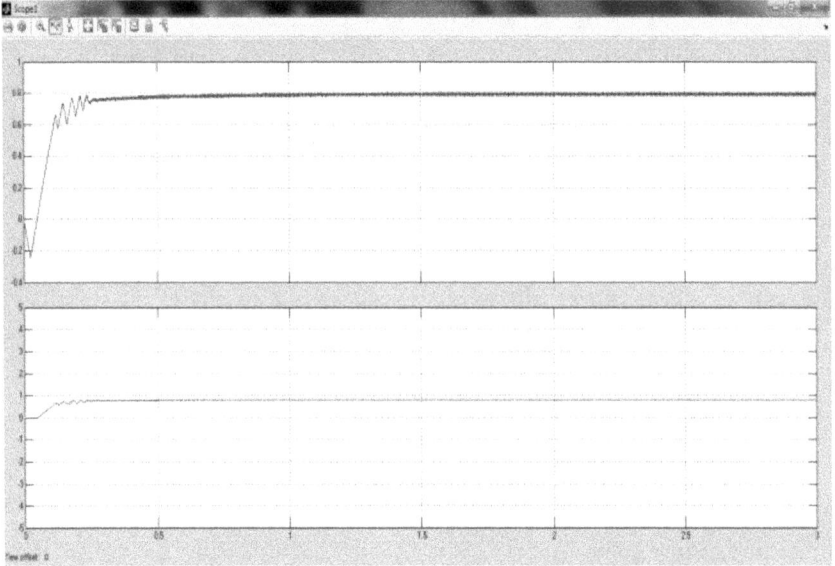

Current Measurement output

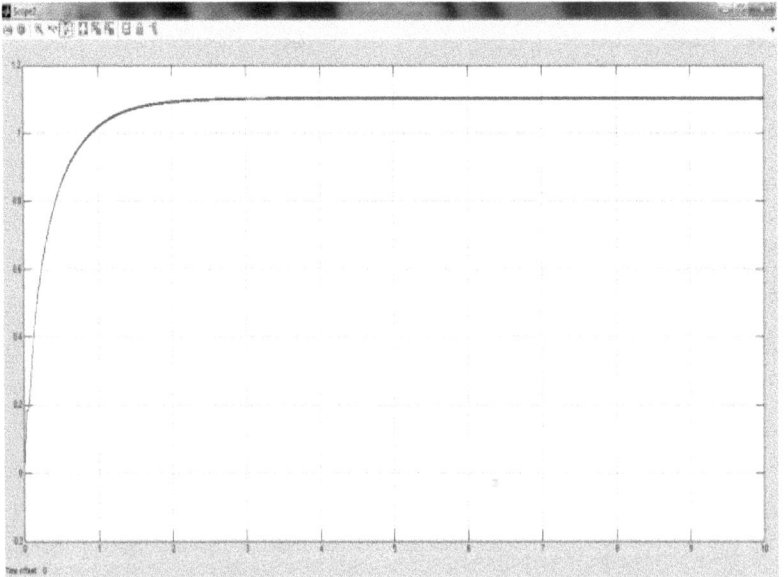

54

Voltage Measurement output

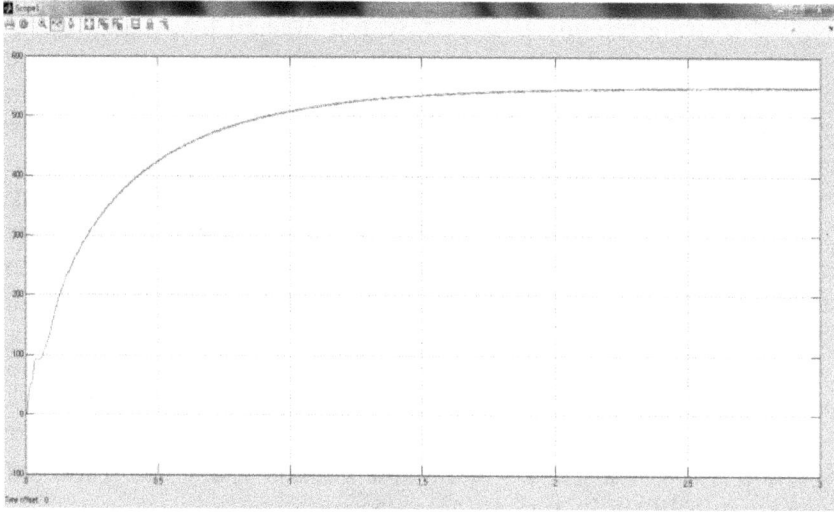

Output

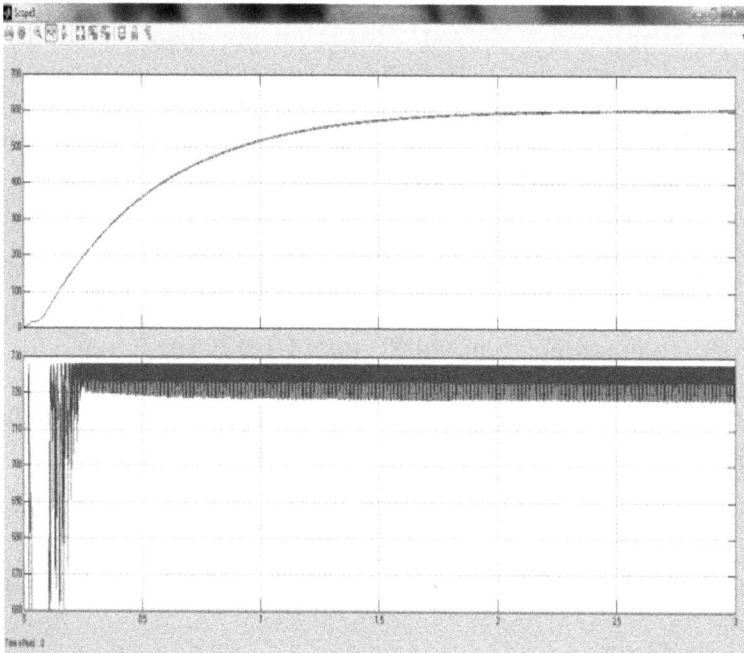

55

6. CONCLUSIONS

Another GMPPT calculation is a somewhat hidden condition, assigned to a more conservative research area to track the highest point in the world. The proposed strategy divides the voltage thread into several small sub-areas and provides some examples. Then, in this respect, the I-V curve of the cluster in each area approximates to another curve in which the current value is more prominent than the actual value of the similar operating voltage. Therefore, considering the examples at the time and the approximate I-V curve, the research field of GMPP is limited. Finally, use P & O calculations to follow the exact GMPP. The technology provided by the replay and exploration results was found in some tests and disputes to follow GMPP in all PS projects. The results of this work were compared with the results of four different strategies, and the highest exposure was confirmed under all test conditions. The provided MPPT method is not sensitive to the module boundary, and can track and display the MPP of PV modules of various shapes.

REFERENCES

[1] Y.-J. Wang and P.-C. Hsu, (2011)"An investigation on partial shading of PV modules with different connection configurations of PV cells," *Energy*, vol. 36, pp. 3069-3078, 2011.
[2] F. Rong, X. Gong, and H. Shoudao,(2017) "A novel grid-connected PV system based on MMC to get the maximum power under partial shading conditions." *IEEE Trans. Power Electron*, vol. 32, no. 6, pp. 4320-4333, June 2017.
[3] I. Abdalla, J. Corda, and L. Zhang (2013), "Multilevel DC-link inverter and control algorithm to overcome the PV partial shading," *IEEE Trans. Power Electron*, vol. 28, pp. 14-18, 2013.
[4] C. Woei-Luen and T. Chung-Ting (2015), "Optimal Balancing Control for Tracking Theoretical Global MPP of Series PV Modules Subject to Partial Shading," *IEEE Trans. Ind. Electron*, vol. 62, pp. 4837-4848, 2015.
[5] K. Sundareswaran, S. Peddapati, and S. Palani (2014), "MPPT of PV systems under partial shaded conditions through a colony of flashing fireflies," *IEEE Trans. Energy Convers*, vol. 29, pp. 463-472, 2014.

[6] K. Sundareswaran, P. Sankar, P. S. R. Nayak, S. P. Simon, and S. Palani (2015), "Enhanced Energy Output From a PV System Under Partial Shaded Conditions Through Artificial Bee Colony," *IEEE Trans. Sustain. Energy*, vol. 6, pp. 198-209, 2015.

[7] B. N. Alajmi, K. H. Ahmed, S. J. Finney, and B. W. Williams (2013), "A Maximum Power Point Tracking Technique for Partially Shaded Photovoltaic Systems in Microgrids," *IEEE Trans. Ind. Electron*, vol. 60, pp. 1596-1606, 2013.

[8] Z. Lin, C. Yan, G. Ke, and J. Fangcheng,(2011) "New Approach for MPPT Control of Photovoltaic System With Mutative-Scale Dual-Carrier Chaotic Search," *IEEE Trans. Power Electron*, vol. 26, pp. 1038-1048, 2011.

[9] S. Lyden and M. E. Haque,(2016) "A simulated annealing global maximum power point tracking approach for PV modules under partial shading conditions," IEEE *Trans. Power Electron*, vol. 31, pp. 4171-4181, 2016.

[10] Pedro Neves, D. Gonçalves, J. G. Pinto, Renato Alves, João L. Afonso (2009) "Single-Phase Shunt Active Filter Interfacing Renewable Energy Sources with the Power Grid", *IEEE Trans.* pp. 3264-3269, .

[11] J. Ahmed and Z. Salam (2015), "An Improved Method to Predict the Position of Maximum Power Point During Partial Shading for PV Arrays," *IEEE Trans. Ind. Inform*, vol. 11, pp. 1378-1387, 2015.

[12] Y. Wang, Y. Li, and X. Ruan (2015), "High Accuracy and Fast Speed MPPT Methods for PV String Under Partially Shaded Conditions," *IEEE Trans. Ind. Electron*, vol. PP, pp. 1-1, .

[13] Y. Hu, W. Cao, J. Wu, B. Ji, and D. Holliday (2014), "Thermography-based virtual MPPT scheme for improving PV energy efficiency under partial shading conditions," *IEEE Trans.* Power Electron, vol. 29, pp. 5667-5672, 2014.

[14] C. Kai, T. Shulin, C. Yuhua, and B. Libing (2014), "An Improved MPPT Controller for Photovoltaic System Under Partial Shading Condition," *IEEE Trans. Sustain. Energy*, vol. 5, pp. 978-985, 2014.

Chapter 5

Accuracy assessment of various supervised classifiers in Land Cover Classification of C-Band Synthetic Aperture Radar (SAR) Sentinel-1Satellite Imagery

Susan John[1] and A. O. Varghese[2]
[1]St. Francis De sales College Seminary Hills, Nagpur, Maharashtra, India
[2]Department of Space, RRSC-Central, NRSC/ISRO, University Campus PO, Amravati Road, Nagpur, Maharashtra, India

1. Introduction

Remote sensing data efficiently gives information about the earth surfaces at low cost and with continuous global coverage thus helps in data exploration and it can be used in emergency management such as flood situation or oil spill in the sea etc.remote sensing also helps in solving many real-life problems such as land resource planning, disaster management, water resources management, forestry, coastal monitoring and land cover and land use etc., in a short period. One of biggest challenge in front of remote sensing community is to accurately classify the remotely sensed data into a thematic map as it is complex in nature and its accuracy depends upon various factors such as properties of sensor used i.e. airborne or space borne, nature of class selected i.e. complexity and heterogeneity of the study area such as presence of mosaic landscapes etc., selection of proper classifier etc. Some of commonly used remotely sensed optical data are Satellite Pour l'Observation de la Terre, or Satellite for observation of Earth (SPOT 5 with 10 meters resolution)[1], Landsat-8 with spatial resolutions 30 meters [2], Advanced Space-borne Thermal Emission and Reflection Radiometer (ASTER- 15 meters spatial resolutions [3], and Moderate Resolution Imaging Spectroradiometer (MODIS- 250 meters for band 1 and 2, 500 meters for bands 3 to 7 and 1000 meters for bands 8 to 36 [4] etc., which differs in resolutions, scales etc., and the primary mission is to obtain the imagery of Earth for various applications such as forestry, geology, water resource management GIS applications etc.

The proper choice of suitable sensor data is needed because for a specific purpose only certain particular remotely sensed data can be used for e.g. A spatial resolution more than 5m gives better accuracy of heterogeneous data for land use and land cover classification mapping[5]. Another important parameter that affects selection of sensor data is the presence of clouds that affects the quality of optically sensed data. Therefore now-a-days optical data is replaced by synthetic aperture radar (SAR) data, which acquires not only high resolution images but also provides continuous data of day and night data of all weather conditions and which are not affected by presence of cloud cover as optical data is affected by presence of cloud [36]. In order to generate two-dimensional images SAR sensors sends their own electromagnetic waves and receives back those signals after being scattered from objects. A high-resolution SAR image can be obtained one perpendicular to the sensor direction i.e. in range direction and other is azimuth direction which is parallel the sensor direction. The various frequencies of electromagnetic (EM) waves emitted by SAR antenna are L-band (1 to 2 GHz), C-band (4 to 8 GHz) and X-band (8 to 12 GHz) [6].

Sentinel-1 is a two satellite constellation which operates on C-Band of microwave region and it to provide C-Band SAR data. This satellite carries single C-band synthetic aperture radar (SAR) instrument operating at a frequency of 5.405 GHz.The C-SAR instrument supports operation in dual polarization (HH+HV, VV+VH) implemented through one transmit which send either H or V (switchable) polarized wave and receive two parallel polarized waves for H and V polarization [37], it has capability to obtain both day & night imagery as well as small movement on the ground thus helps in continuous monitoring of land and sea. It can be freely downloadable form:scihub.copernicus.eu.

According to Lu& Weng, apart from the imagery appropriateness, use of correct classification method is also needed which affects the results of land cover mapping[7]. Hence another factor that affects the accuracy of image classification is the proper choice of image classification algorithms. Image classification is divided into two i.e. supervised classification and unsupervised classification. Supervised classification requires prior knowledge of classes, as it teaches the classifier to determine decision boundary in feature space by training samples in the form of pixel.

60

In this technique the classification result is compared with the ground truth. Finally the accuracy is calculated by computing the confusion matrix and the result is compared with the ground truth i.e. the accuracy of the decision boundary depends upon the number of training sample and type of classes [8].

Many classifiers have been developed and tested on remotely sensed data for better interpretation.

As per Szusteret al., for the tropical coastal zone SVM showed highest accuracy as compared to the neural network techniques and maximum likelihood classifier[9]. But maximum likelihood approach faces the disadvantage of overestimating the class values even though it shows better accuracy when compared to Mahalanobis distance and Minimum Distance algorithms [10]. Accuracy of SVM dependent on inclusion of appropriate training data set and hence large number of training sets are needed[11]. Parallelepiped classifier performs well and it requires a supervised training by an experienced analyst[12] [13]. Neural Net faces serious disadvantage of misclassification which further affects its performance[14]. Apart from these issues other issues such as mixed pixel problems, overlapping classes still remains and have not been solved completely. These various problems have driven many researches for improving the various algorithms so as to increase the accuracy level to predicate the feature correctly.

Here attention is focused on performances of supervised classifiers. The overall objective of this study is to evaluate the performances of seven supervised classifiers namely Parallelepiped Classification, Minimum Distance Classification, Mahalanobis Distance Classification, Maximum Likelihood Classification, Support Vector Machine classification, Neural Net Classification and Binary encoding Classification on freely available Sentinel-1 satellite imagery. As this type of C-band satellite imagery is new, more researches is necessary to conduct and to evaluate the usefulness of this imagery, as only few studies, are available in literature for land use & cover mapping. The results from this study can provide insights into proper selection of classifier in classifying Sentinel-1 C band remotely satellite data for land cover classification.

2. Study area and Data set

2.1 Data and data preparation: The study area is situated in Nagpur district of Maharashtra state, in central part of India. The image data products being used in this study are of SAR C band Sentinel-1 satellite data captured on 30-05- 2020, in IW mode and swath, with polarization VV-VH. Product type: GRD used in this study. The satellite data were procured from http//scihub.copernicus.eu. The study was conducted on freely available satellite imagery Sentinel-2 using ENVI PRO software.

2.2 Study area: The test site is located in central India, Nagpur, Maharashtra, India; its geographical location co-ordinates of the polygon selected are longitude-80.33 latitude-21.05, longitude-80.65 latitude-22.56, longitude-78.22 latitude- 22.98, longitude-77.93 latitude-21.48, longitude-80.33 latitude- 21.05.

2.3 Land cover detection and analysis: To work out the land cover classification on sentinel-1 C-band SAR satellite data classification was accomplished using ENVI software. With the help of Google earth, ground verification was done for doubtful areas. Five land cover types were identified for the study area: (1) waterbodies, (2) cropland(3) forest, (4)Scrub land and (5) built-up area. Nearly 50 training areas were selected for each class using region of interest in the ENVI PROsoftware all the classifications were conducted by these training sites. For accuracy assessment, ground truth information collected from the field was used.

3. Proposed Method

We aimed at evaluating the performance of seven classifier algorithms on Sentinel-1 C-band SAR image. In the following subsections, a brief explanation of the algorithms is provided.

a) Maximum likelihood

Maximum Likelihood (ML) is a supervised classification method derived from the Bayes theorem, which assumes that the statistics for each class in each band are normally distributed and calculates the probability that a given pixel belongs to a specific class. And also assumes that the distribution of the data within a given class obeys a multivariate Gaussian distribution [15]. Firstly different types of land cover are determined in the study area followed by the training pixels for each of the desired classes by determining the Jeffries-Matusita (JM) distance which is measure of class separability of the training pixels. Then the mean vector and covariance matrix of each class are estimated followed by classification of every pixel. Thus a probability threshold is selected, which determines the classification of pixels to a particular class. If the selected pixel is equal to probability threshold then it is assigned a class otherwise the pixels remains unclassified. The Maximum likelihood function describes ellipsoidal, i.e. equi-probability contours are decision boundaries and its shape depends upon the orientation and relative dimensions of the axes of the ellipse [15], to represent feature space of the pattern of pixels belonging to a given class.

b) Support Vector Machine

In a binary classification for linearly separable classes, SVM maximizes the distance from the data points of each class thus helps to create the optimal separating linear hyperplane from each variable. Any training data points on the two hyperplanes viz, parallel and either side to optimum hyperplane, are termed as support vectors,[16]. This optimum hyperplane not only separates the training data set with a maximal margin. And for classes which are not linearly separable, the training data cannot be separated without error, hence in such case the training set should be separated with minimal number of errors by mapping the input data onto a higher dimensional space called feature space or Hilbert space which is dot product space. Thus SVM is optimized using a kernel function such as linear, polynomial, radial basis function(RBF)& sigmoid in order to search hyperplane in a multidimensional feature space. Some studies [17]conclude that the most commonly used kernels functions are the non-linear polynomial and radial basis kernels. So far RBF kernel is the best choice for practical applications[18]. Thus the SVM with RBF kernels is used in this study.

c) Parallelepiped Classifier:

It is also known as box classifier. Firstly a threshold of each class signature is decided by an analyst i.e. analyst is directly involved in the training sites selection. For each band in the multispectral images, a valid range of intensity values is specified by a minimum and maximum value which resembles a box. Hence the parallelepiped classification makes use of a rectangular box whose boundaries are defined by maximum i.e. high threshold and minimum i.e. low threshold pixel values of each bands [39]., which is fitted for each class. Hence the position of pixels determines the classification, if the pixel value falls inside the rectangular box then it is classified and hence it is assigned class otherwise it is unclassified. There arises one more class called overlap class, because of those pixels which falls in more than one class and is caused due to high covariance between the bands [19].

d) Neural Net Classifier:

Neural Network exactly replicates the mechanism of human brain, where the intelligence is stored not only in neural pathways but also in the memory. In this, knowledge is applied in the forms of weights applied to a node hence multiplicative values are applied to an input. A supervised network is presented so that the Neural Net learn itself, like human being learns by experience, by setting weights which will produce a specified output. When the new data is presented, neural net applies the weight, and the output generated after comparing with the previous experience. There are many types of neural networks which can be used for classification purposes, most widely used is multilayer perceptron's (MLP). It faces serious disadvantages such as large computation is needed, requires greater training time etc. [20]. Hence Neural Net first trains a small portion of randomly selected pixels from a specific image under study which is net and then further it is applied to remaining part of the image. In order to optimize the accuracy of neural net, a good and proper selection of network design is needed which will automatically reduce the training time [20]. As it faces wide range of factors that limit the use of Neural Net [21]. Another commonly used algorithm is the back-propagation algorithm is supervised training algorithm, is based on minimization of the error between the actual network outputs and the outputs of training input or output pairs [22].Thus as a result hence error is propagated back to the input layer from the output. Thus, it helps in renewing the weights of the backward path i.e. it is based on trial and error process of changes of model parameters.

64

e) Mahalanobis distance Classifier:

It is a parametric classifier, which is based on variance-covariance matrix. Mahalanobis distance classification is same as maximum likelihood classification. Based on the user-defined classes, training data are given to specify the spectral classes. When the covariance matrix is the unit matrix the Mahalanobis distance equals to Euclidean distance (ED). Smaller the value of Mahalanobis distance gives greater chances of pixels being correctly classified[23].

f) Minimum Distance Classifier:

This classification based on matching of previously stored template i.e. template matching or pattern recognition, in which a matching of previously stored template, based on the some previously defined measure of similarity, for each class to an unknown pattern is done after matching the classification . Here both templates and unknown patterns are distribution functions which are measure of distance between distributions function i.e. measure of similarity. The image is classified according to the shortest distance class i.e. straight Euclidean Distance. According to [24] the distribution functions are a set of random measurement vectors from each distribution of interest and classification is based on estimated distributions not on actual distribution. Measuring the distance whose distribution function is nearest to the unknown distribution classifies the distribution into a class. The minimum distance indicates maximum similarity. It is performed very well when the number of training samples per class is limited

g) Binary Encoding Classifier:

The basic idea of binary encoding is to reduce the large amount of data without altering the information as much as possible [25] by reducing the information of a pixel which is often represented as 8 bit per channel into one or two bits per channel only i.e. individual pixels are considered [26]. In this [27] mean grey values over all available channels i.e. spectral mean values are calculated from the individual channel grey values [28].

4. Results and Discussion

In order to evaluate the classifier performance, the assessment methods plays a vital role which are divided into three process such as training the data set using input pattern classifier fits the training data and to predict class labels for unseen data, since the class labels of testing samples are unknown, in order to evaluate the performances of the trained model, validation is done which provide an unbiased evaluation of the trained model. And last process is the testing its accuracy in terms of percentage.

The study area involved the number of classes such as waterbody, cropland, forest, scrubland and built-up area i.e. it is multi-class classification problem not the binary classification which involves only two classes.

The classification model was trained to predict the true classes of unknown data set, in training phase.

This classification model produces outputs by generating confusion matrix or a contingency table. The left to right- diagonal elements represents correct predictions and the rest are the incorrect predictions.

The confusion matrix or error matrix is a square matrix that contains statistics for assessing accuracy (%) by showing the degree of misclassification among the different feature class which compares relation between classification result and reference data on a class by class basis [29,30,31&32]. Information of producer accuracy, user accuracy and overall accuracy can be obtained from confusion matrix. Producer accuracy represents the percentage of a correctly classified ground class whereas user accuracy is the percentage of correctly classified pixels.

The off-main diagonal cell of confusion matrix represents the number of misclassified pixels in the form errors of omission and errors of commission. The overall accuracy is obtained by dividing the sum of main diagonal entries of the error matrix by the total number of samples. User accuracy and producer's accuracy to commission and omission relations are given below. The relationships are calculated as

66

User's accuracy (reliability) = 100% - commission error (%) and

Producer's accuracy = 100% - error of omission (%)

The user's accuracy is the ratio between the number of correctly classified and the row total and producer's accuracy is ratio of between the number of correctly classified and the column total. Two important measures of accuracy are user's accuracy and user's accuracy.

From the analysis of producer accuracy (%), neural net classifier predicted correctly 88.69% of water body followed by maximum likelihood classifier with 85.07 % and SVM with 80.09 %, whereas Mahalanobis classifier and binary classifier predicated only 50.23% and 40.27 respectively. For cropland, binary classifier reported highest producer accuracy with 98.09%, followed by neural net with 94.9%, parallelepiped with 94.1% and SVM reported 93.78% whereas maximum likelihood classifier reported lowest producer accuracy of 89.79%. For forest highest producer accuracy were reported for neural net with 99.85%, followed by SVM with 99.85% and maximum likelihood classifier with 99.56% and lowest for minimum distance classifier with 75.48 %. For scrubland, maximum likelihood classifier reported 88.85% of producer accuracy followed by SVM of 79.14% and neural net with 76.26% and lowest producer accuracies were predicated for Mahalanobis classifier with 67.63% and binary classifier failed totally. And for built-up area SVM reported highest producer accuracy with 97.76%, then maximum likelihood classifier with 96.13% and neural net classifier with 96.26% and lowest for parallelepiped classifier with 53.87% and here again binary classifier reported zero.

Highest overall accuracy and kappa coefficient was obtained for support vector machine classifier with 94.66% & 0.93 respectively followed by neural net with overall accuracy 94.53% & kappa coefficient 0.92 and maximum likelihood classifier with overall accuracy 94.26% & kappa coefficient equals 0.92 and lowest accuracy was reported for binary encoding classifier with overall accuracy 42.10%& kappa coefficient equals 0.28.

Binary classifier reported the lowest accuracy for all classes, except for forest which correctly classifier 99.7% and for cropland 98.07%.

It failed to classify scrubland and built-up area completely.though binary encoding classifier is very simple and effective classifier, but it requires complete labeling of all classes [33], which hinder its performances. It efficiency also gets affected in dealing with hyper spectral data. The parallelepiped classifier only reported highest accuracy for cropland & forest and for all other class it reported medium and for scrub-land it reported lowest. Size of cropland area & forest is large over the whole imagery and its distribution is also normal hence parallelepiped gave highest producer accuracy for cropland. It is also noticed that the overall accuracy depends upon the distribution of training sites in the imagery and the selection of training sites also. And another drawback is it depends on how well the analyst selects training data i.e. it requires a supervised training. The accuracy of this classification technique is affected by presence of overlapping classes and also depends on pixel value, if it is below the threshold it will remain unclassified.presence of mixed pixels greatly affected the accuracy of Mahalanobis classifier and minimum distance classifier. Maximum likelihood classifier also faces disadvantage of over-prediction of the positive classes. The accuracy of Maximum likelihood classifier is affected by presence of mixed pixels and also affected by overlapping classes. As all the five classes is partially spectrally overlapping and there is also presence of mixed pixels which affects its accuracy. And also, it is noticed that for the spectrally overlapping class, it favors the dominant class but it also causes over fitting. As according to [34], if spectral overlapping between the classes is high Maximum likelihood classifier gives a noisy classification, because of isolated labeled pixels inside patches of other classes gives noisy impression. Here in this case the spectral overlapping between the class is less and its distribution is normal Gaussian distributed hence Maximum likelihood classifier gave good accuracy with 94.26 % overall accuracy & kappa coefficient 0.92 respectively, same is also reported in [35].Neural net classifier also correctly classified all classes but its prediction was affected because of presence of mixed pixels and hence resulted into more misclassification.The accuracy obtained from SVM is not highest except for built-up area and forest because appropriate support vectors are not obtained hence a greater number of training sites must be included. Hence its accuracy is depending on size of training data SVM handles the mixed pixels problems hence it gave highest producer and user accuracy. Performance of neural net classifier and SVM classifier depends upon effective use of user-defined parameters.

SVM, neural net and maximum likelihood showed relatively same accuracy.classified maps of the study are were generated by using all seven classifiers as shown in fig. no. 3 and fig. no.4 respectively. Results of classifications from all the classifiers have been verified by ground observations using Google Earth Pro.Further from visual analysis also SVM classifier showed highest accuracy in classifying the land cover.

5. Conclusions

The present study examined the accuracy of seven different supervised classifiers for land cover classification of the study area, part of central Indian region in Nagpur. Results of classifications have been verified by ground observations using Google World View and ground truth collected. Overall, SVM classifier produces a more sensible and realistic results compared to other classifiers. From the analysis of result, it is observed that SVM classifier has reported the highest overall accuracy and kappa coefficient 94.66% and 0.93, followed by neural net 94.53% and 0.92 and binary classifier has the least accuracy of 42.10% & 0.28 respectively. Forest was accurately classified by SVM and Neural netclassifier 99.85%.However in comparison to neural network, SVM classified all classes more accuratelyand number of misclassified pixels is less than neural network. Binary encoding classifier failed totally in classifying the scrubland and built-up areaseparately. Hence SVM classifier because of its high computational efficiency, it is considered to be very good classifier, in classifying Sentinel-1 C-band synthetic aperture radar (SAR) image as compared to other classifiers.

Acknowledgements

The first author extend her sincere gratitude to Regional Remote Sensing Centre-Central (RRSC-Central), National Remote Sensing Centre (NRSC) for providing the opportunity and facility to conduct the study. The technical support provided by RRSC-Central staff is duly acknowledged. We are also thankful to USGS Earth Explorer maintained by the NASA and Bhuvan-Indian Geo Platform maintained by NRSC, ISRO for online accessibility of data.

References

[1] https://earth.esa.int/web/eoportal/satellite-missions/s/spot-5

[2]https://www.usgs.gov/faqs/what-are-band-designations-landsat-satellites?qt-news_science_products=0#qt-news_science_products

[3]https://lpdaac.usgs.gov/data/get-started-data/collection-overview/missions/aster-overview/

[4]https://ladsweb.modaps.eosdis.nasa.gov/missions-and-measurements/modis/

[5] R. Welch (1982) Spatial resolution requirements for urban studies, International Journal of Remote Sensing, 3:2, 139-146, DOI: 10.1080/01431168208948387

[6]https://www.esa.int/Applications/Telecommunications_Integrated_Applications/Satellite_frequency_bands

[7] LuD.& Weng Q. (2007): A survey of image classification methods and techniques for improving classification performance, *International Journal of Remote Sensing*, 28:5, 823-870.

8] Mather P. and Tso B. (2010), Classification methods for remotely sensed data. *CRC press*.

[9] Szuster B.W., Chen Q. and Borger M.(2011), A comparison of classification techniques to support land cover and land use analysis in tropical coastal zones. *Applied Geography*, 31, 525–532.

[10] Al-Ahmadi F. S. And Hames A. S. (2009), Comparison Of Four Classification Methods To Extract Land Use And Land Cover From Raw Satellite Images For Some Remote Arid Areas, Kingdom Of Saudi Arabia; *Earth Sci.* 20(1) ,167–191.

[11] Foody G.M.& Mathur A. (2004). A relative evaluation of multiclass image classification by support vector machines. *IEEE Transactions on Geoscience and Remote Sensing* 42, 1335–1343.

[12] Xiang M., HungC.C., PhamM., Bor-Chen K. and ColemanT. (2005), "A parallelepiped multispectral image classifier using genetic algorithms," Proceedings. *2005 IEEE International Geoscience and Remote Sensing Symposium*, 2005. IGARSS '05., 2005, pp. 4 pp.-, doi: 10.1109/IGARSS.2005.1526216.

[13] Swain, P. H. and Davis S. M. (1978), *Remote Sensing: The Quantitative Approach*, New York, NY: McGraw-Hill.

[14] VictorB. L. and ZhangG. P (1999)., "The effect of misclassification costs on neural network classifiers," *Decision Sci.*, vol. 30, pp. 659–682, 1999.

[15] TSO B. and MATHER P.M., (2001), Classification Methods for Remotely Sensed Data (New York: Taylor and Francis Inc).

[16] Mathur A. & Foody G.M. (2008). Multiclass and Binary SVM Classification: Implications for Training and Classification Users. *Geoscience and Remote Sensing Letters, IEEE.* 5. 241 - 245. 10.1109/LGRS.2008.915597.

[17] Huang C., DavisL. S., and TownshendJ. R. G. (2002). "An Assessment of Support Vector Machines for Land Cover Classification." *International Journal of Remote Sensing* 23 (4): 725–749. doi:10.1080/01431160110040323.

[18]PatleA. and ChouhanD. S., "SVM kernel functions for classification," *(2013) International Conference on Advances in Technology and Engineering (ICATE),* 2013, pp. 1-9, doi: 10.1109/ICAdTE.2013.6524743.

[19]Perumal And BhaskaranR., (2010) "Supervise Classification Performance Of Multispectral Images", Journal Of Computing, Volume 2, PP 124-129.

[20] Miller D. M., Kaminsky E. J., Rana S. (1995). Neural Network Classification of Remote Sensing Data. *Computers & Geosciences,* 21(3), 377-386.

[21]Kavzoglu T., (2001), An investigation of the design and use of feed-forward artificial neural networks in the classification of remotely sensed images. PhD thesis, School of Geography, The University of Nottingham, Nottingham, UK.

[22] Yilmaz I.; Kaynar O. (2011),Multiple regression, ANN (RBF, MLP) and ANFIS models for prediction of swell potential of clayey soils. *Expert Syst. Appl.* 38, 5958–5966.

[23] Talukdar S., Singha, P.Mahato, S., Shahfahad, Pal S., Liou Y. & Rahman A (2020). Land-Use Land-Cover Classification by Machine Learning Classifiers for Satellite Observations—A Review.*Remote Sensing.* 12. 10.3390/rs12071135

[24] Wacker A. G. And Landgrebe D. A., "Minimum Distance Classification In Remote Sensing" (1972*).LARS Technical Reports.*Paper 25.

[25] Jia X. and Richards J. A. (1993), Binary Coding of Imaging Spectrometer Data for Fast Spectral Matching and Classification, *REMOTE SENS. ENVIRON.* 43:47-53.

[26] Xie H., Heipke C., Lohmann P., Soergel U., Tong X., & Shi W. (2011). A New Binary Encoding Algorithm for the Simultaneous Region-based Classification of Hyperspectral Data and Digital Surface Models. Photogrammetrie - Fernerkundung - *Geoinformation,* 2011(1), 17–33.

[27] MazerA.S., MartinM., LeeM., and SolomonI.E.,(1988) "Image processing software for imaging spectrometry data analysis", *Remote Sensing of Environment*, 24, pp. 201-210, 1988.

[28] Wei L., Chen Chen, Hongjun S., and Qian D., (2015)"Local Binary Patterns and Extreme Learning Machine for Hyperspectral Imagery Classification," *IEEE Transaction on Geo Science and Remote Sensing*, vol. 53, no. 7, pp. 3681-3693, 2015

[29] Smits P.C., Dellepiane S.G. And Schowengerdt R.A., (1999), Quality assessment of image classification algorithms for land-cover mapping: a review and a proposal for a cost based approach. *International Journal of Remote Sensing*, 20, pp. 1461–1486.

[30] Congalton R.G. And Plourde L., (2002), Quality assurance and accuracy assessment of information derived from remotely sensed data. In J. Bossler (Ed.), Manual of Geospatial Science and Technology (London: Taylor & Francis), pp. 349–361.

[31] Foody, G.M., (2002b), Status of land cover classification accuracy assessment. *Remote Sensing of Environment*, 80, pp. 185–201.

[32] Foody, G.M., (2004a), Thematic map comparison: evaluating the statistical significance of differences in classification accuracy. *Photogrammetric Engineering and Remote Sensing*, 70, pp. 627–633.

[33] Munoz-M.J., BruzzoneL., and G. Camps-Valls. (2007). "A Support Vector Domain Description Approach to Supervised Classification of Remote Sensing Images." *IEEE Transactions on Geoscience and Remote Sensing 45 (8):* 2683–2692. doi:10.1109/TGRS.2007.897425.

[34]Cortijo, F.J., (1995), A Comparative Study of Classification Methods for Multispectral Images. PhD thesis, DECSAI, Universidad de Granada,   Spain.Available   at http://decsai.ugr.es/~cb/entrada_tesis.html, electronic edition.

[35] Huang C., DavisL. S. and TownshendJ. R. G.(2002). "An Assessment of Support Vector Machines for Land Cover Classification." International Journal of Remote Sensing 23 (4): 725–749. doi:10.1080/01431160110040323.

[36] https://www.mdpi.com/1424-8220/15/10/25366/pdf

[37] https://sentinels.copernicus.eu/web/sentinel/missions/

[38] https://www.l3harrisgeospatial.com/docs/maximumlikelihood

[39]http://ecoursesonline.iasri.res.in/mod/page/view.php?id=20

Decomposition	Maximum Likelihood Classifier					Support Vetor Machine					Parallelepiped Classifier				
Land Cover Class	1	2	3	4	5	1	2	3	4	5	1	2	3	4	
1 Waterbody	85.07	0	0	0	0.27	80.09	0	0	0	0	73.76	0	0	0	5.1
2 Cropland	0	89.79	0	2.88	0	0.9	93.78	0	7.55	0	0	94.1	4.28	5.76	0
3 Forest	0.45	2.07	99.56	8.27	2.45	0.45	1.59	99.85	13.31	2.04	1.81	5.26	94.83	52.88	10.87
4 Scrub Land	8.14	8.13	0.3	88.85	1.15	5.88	4.63	0	79.14	0.2	15.38	0.48	0.59	41.37	29.08
5 Built-Up Area	6.33	0	0.15	0	96.13	12.67	0	0.15	0	97.76	2.71	0	0	0	53.87
	Overall Accuracy 94.26% ; Kappa Coefficient 0.92					Overall Accuracy 94.66%; Kappa Coefficient 0.93					Overall Accuracy 70.32% ;Kappa Coefficient 0.61				

Decomposition	Binary Encoding Classifier					Neural Net Classifier					Mahalanobis Distance Classifier				
Land Cover Class	1	2	3	4	5	1	2	3	4	5	1	2	3	4	
1 Waterbody	40.27	0	0	0	0	88.69	0	0	0	0	50.23	0	0	0	0
2 Cropland	58.82	98.09	0.3	89.93	1.77	0.45	94.9	0	10.07	0	2.26	91.23	0	17.63	0
3 Forest	0.9	1.91	99.7	10.07	98.23	0.45	1.44	99.85	13.67	2.65	0	2.39	86.26	8.63	8.97
4 Scrub Land	0	0	0	0	0	4.52	3.67	0	76.26	1.09	38.01	6.98	5.32	67.63	2.92
5 Built-Up Area	0	0	0	0	0	5.88	0	0.15	0	96.26	9.5	0	8.42	6.12	88.11
	Overall Accuracy 42.10%; Kappa Coefficient 0.28					Overall Accuracy 94.53%; Kappa Coefficient 0.92					Overall Accuracy 84.03%; Kappa Coefficient 0.77				

Decomposition	Minimum Distance Classifier				
Land Cover Class	1	2	3	4	5
1 Waterbody	59.73	0	0	0	0.34
2 Cropland	0.45	92.82	2.95	4.68	0
3 Forest	0.9	4.15	75.48	19.78	2.04
4 Scrub Land	3.62	3.03	20.09	69.42	4.55
5 Built-Up Area	35.29	0	1.48	6.12	93.07
	Overall Accuracy 85.13% ; Kappa Coefficient 0.79				

Table 1 Confusion matrices of classifications based on different classifiers

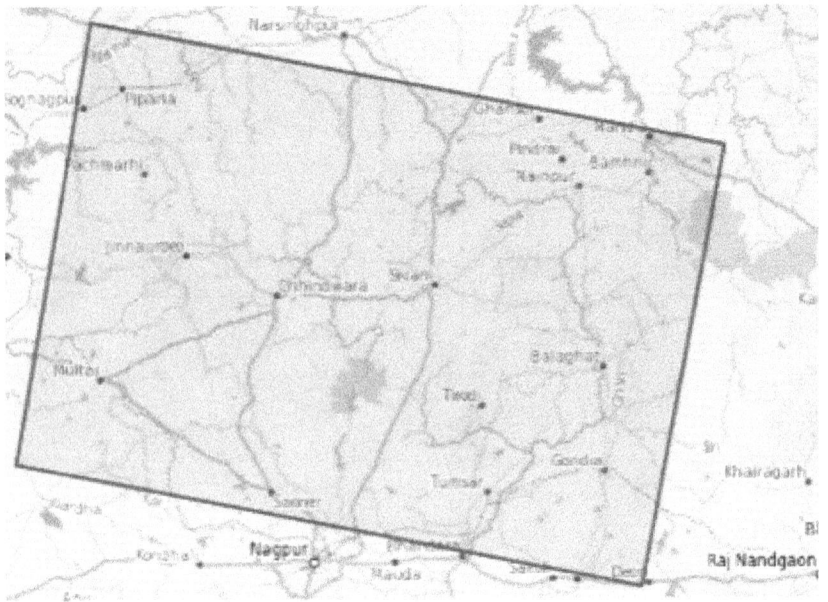

Figure 1: Image of Nagpur, Maharashtra

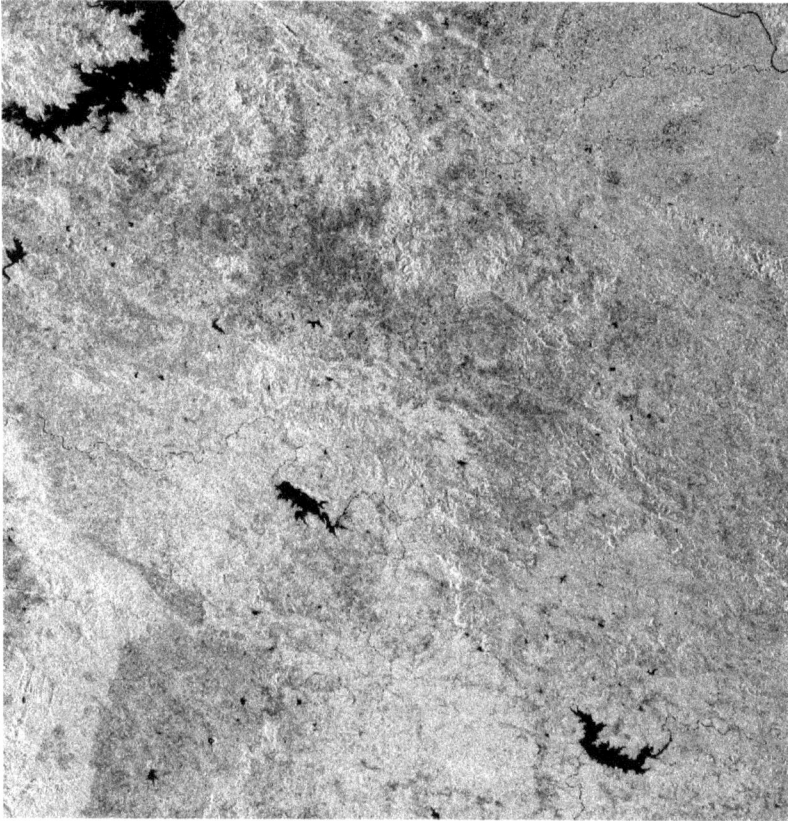

Figure 2: Sentinel-1 C-band synthetic aperture radar (SAR) image of Nagpur, Maharashtra

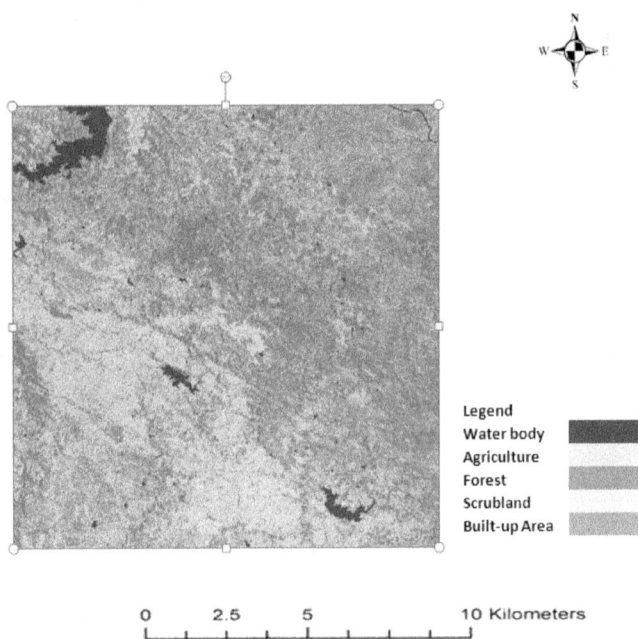

Fig 3: Classified output Sentinel-1 C-band synthetic aperture radar image of a portion of Nagpur, Maharashtra using SVM classifier.

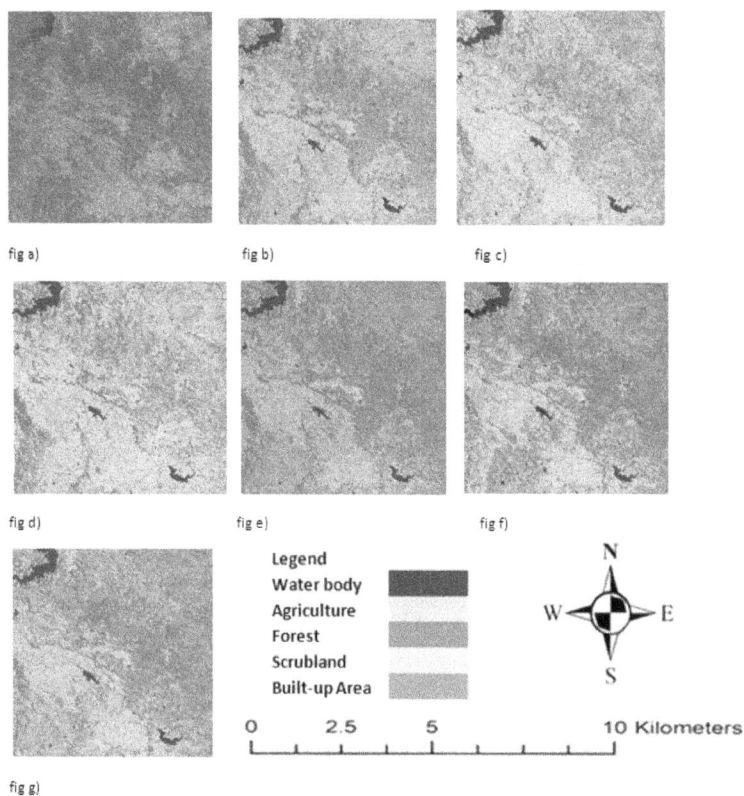

Fig 4 a) classified image using Binary Encoding Classifier; b) classified image using Mahalanobis Classifier; c)classified image using Maximum Likelihood Classifier; d)classified image using Minimum Distance Classifier; e)classified image using Neural Network Classifier; f)classified image using Parallelepiped Classifier ; g) classified image using Support Vector Machine (SVM) Classifier.

Chapter 6

Application of concepts from Artificial Intelligence to Advanced Manufacturing Systems

E. Fantin Irudaya Raj
Department of Electrical and Electronics Engineering, Dr Sivanthi Aditanar College of Engineering, Tamilnadu, India

An artificial intelligent technique is widely used in several engineering fields. In manufacturing systems, this advanced concept is utilized to increase the overall performance of the system. This book chapter discusses the features, benefits of utilizing artificial intelligence in the manufacturing system. The modern artificial intelligence techniques are adapted in additive manufacturing, flexible manufacturing systems, scheduling operations, and production planning and control techniques. Additive manufacturing is the innovative advanced manufacturing method to produce each part with unique features. The improvement in the additive manufacturing system generates the revolution of Industry 4.0 in the manufacturing system. To improve the quality of the product, reduce the mass of the components, better shape to withstand several boundary conditions, the physical characteristics are effectively optimized by the implementation of artificial intelligence in the additive manufacturing field. The resource allocation, onsite demand service, defect-free product, production capacity of the machines are customized by these advanced techniques. A flexible manufacturing system is an innovative technique for the effective utilization of man-machine resources. The artificial intelligence techniques are adapted to minimize the complexity of decision making in the dynamic situations of scheduling based on the job orders. Also, the CNC machining parameters are optimized by this technique for uplifted production capacity. The several factors affecting the machining production capacity and quality are properly distinguished and controlled to manufacture high-quality products. The production planning and control operations are made by the artificial intelligence methods by the past history knowledge. The complex algorithm with subroutines is eliminated by the simple available previous production schedule. The production plan is further fine-tuned by the learning techniques in artificial intelligence.

1.1 Introduction

The manufacturing industry is a foundation stone of a country economy, livelihood of people and national security. The manufacturing industry is growing with new technological and industrial techniques nowadays. Autonomous robots, internet services, automated guided vehicles, computer numerical control machines, artificial intelligence techniques, sensors, modern software tools play a vital role in this development. The sensors fit in all manufacturing systems to monitor the real-time information from every unit. The wireless data transfer devices collect and transfer several information to the central computer servers. The computer server receives and stores all the information from several wireless nodes for management purposes. The simulation methods [1-9] produce numerical result. The experimental procedure is costlier while running several trials.

The internet with artificial intelligence techniques is utilized for analyzing several pieces of information from various information-gathering systems. The fusion of data from several sensor nodes for data collection is done by an artificial intelligence system. The manufacturing process, material technology and energy consumption are properly optimized by this modern technique. Data mining by modern information technology tools is the next level game-changer tool of several manufacturing methods and manufacturing models. With the current network connectivity speed, the cyber-physical systems, big data analytics the industrial revolution is done. Further, the evaluation of smart transportation, smart city and autonomous vehicle concept takes the manufacturing system to another level of technological development. Artificial intelligence technology is also adapted in the electrical industry [10-22].

In [23] intelligent manufacturing is upgraded to a smart manufacturing concept with the assistance of the artificial intelligence concept. The usage of the internet of things, cyber-physical concepts and central cloud computing techniques gave an impact on the industrial revolution 4.0. Artificial intelligence is not only utilized in the engineering fields but in the medical field also this concept is implemented to analyze the COVID-19 [24-27] affected peoples to better know the pattern of COVID virus spread in the society.

In this book chapter, the utilization of artificial intelligence in the various manufacturing field is clearly denoted. The manufacturing systems have production operations, process scheduling process, production planning process, automation and cost estimation techniques. the requirement of artificial intelligence techniques for various manufacturing fields is discussed. The traditional techniques difficulties and the feasibility of upgrading the field to advance technology by artificial intelligence are explained in an elaborated manner.

1.2 Manufacturing planning and control by artificial intelligence techniques

The manufacturing planning and control methods are mainly implemented in managerial decision-making operations. This concept has three different steps such as inventory planning, manufacturing operation planning and material flow planning [28]. Customer demand is one of the highly varying parameters day-to-day. This dynamic characteristic of job order is effectively handled by several strategic techniques. The inventory is properly managed for the available job orders, the machine is utilized up to maximum level and the work is to be completed in stipulated time by proper manufacturing planning techniques. The main objective of the manufacturing planning method is to manage the job order at the strategic level, material stock for the machining operations at the tactical level and machine availability at the operational level. The computer-aided manufacturing control operations are one of the feasible techniques to obtain the required productivity goals of the manufacturing sectors. The complete automation of several manufacturing planning operations complicates the system. The machine, material and other handling devices are dependent on each other and so the real-time decisions only give better results at different levels of manufacturing operations. Owing to the dynamic type of this problem, the decision making is time-based and the common solution is unstructured. The boundary conditions are continually reviewed for every timely decision making in the systems. Conventional manufacturing practices follow computer-aided production planning operations and shop floor controllers. But these systems are adapted to a remarkable level. The multi-objective problems require more time to complete since several complex variables are to be solved. Further monitoring and controlling the solutions to complex problems

81

is more difficult. The dynamic nature of the problems handled by the human operators is also tough. This main drawback is overcome by implementing artificial techniques in manufacturing planning and control operations [29-34]. Fig. 1 shows the Manufacturing Planning and control process flow chart.

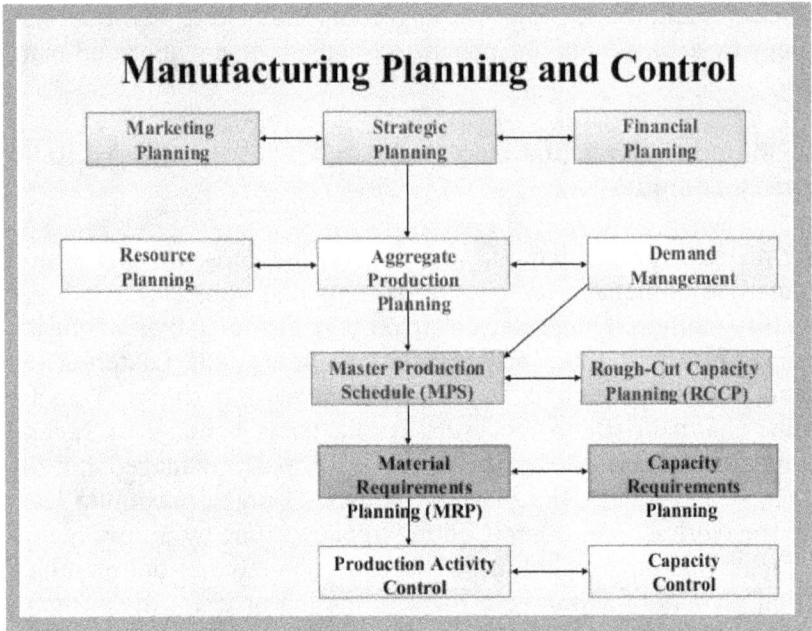

Manufacturing Planning and Control

| Marketing Planning | → | Strategic Planning | ← | Financial Planning |
| Resource Planning | ← | Aggregate Production Planning | ← | Demand Management |

Master Production Schedule (MPS) ← Rough-Cut Capacity Planning (RCCP)

| Material Requirements Planning (MRP) | ← | Capacity Requirements Planning |
| Production Activity Control | → | Capacity Control |

Figure 1. Manufacturing Planning and control process

The computer analytical techniques can provide several logical methods to solve complicated problems and human operators choose the optimum planning charts on the computer systems. The artificial intelligence methods give the most efficient planning methods for enhanced productivity of automated manufacturing systems. Here the human intelligence for decision making in dynamic conditions is transferred to the computer for real-time solutions. Most of the decisions are taken automatically by the artificial intelligence techniques where the human operators take too much time in those situations. Sometimes the information is given to the human operators for decision making to minimize the time taken by the human operators [34]. Thus, this modern artificial intelligence method is the innovative concept to solve several manufacturing planning problems methodically based on the knowledge source from the past history data available.

The implementation of artificial intelligence in all three steps is efficient to schedule various production planning processes. The utilization of capitals such as machine, man, tools are effectively scheduled for the available job orders. Even though the job order is changing from time to time, modern artificial techniques guide several production planning activities for maximum utilization of our resources. The overall performance of the production units could not affect this effective scheduling [35]. For example, the best production plan is done to meet available job orders. The inventory is properly maintained with sufficient raw materials, tools to handle the job orders and the excess amount of materials is avoided based on the past history information. The capital invested for the inventory is optimized for coordination with the available machine capacity. The human and machine utility are balanced according to the demand of the production needs. In the high automation industry, a minimum human workforce is available. In this case, dynamic decision making is difficult due to the lack of a workforce. The artificial intelligence method is the best technique in the highly automated industry, where the automated production plan is created or real-time guidance is given to do the production plan at a minimum time schedule.

1.3 Flexible manufacturing systems scheduling by artificial intelligence method

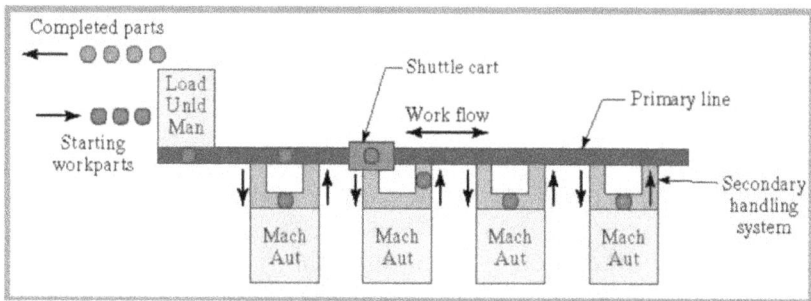

Figure 2. Flexible manufacturing system

In a flexible manufacturing system, each manufacturing unit acts as a separate modular structure. The different types of robots, automatic guided vehicles, autonomous material handler systems and computer-centred machines are available in flexible manufacturing systems. Job scheduling is one of the important tasks in this concept [36]. Each machine is allocated with the job operations. The scheduling is carefully done for maximum machine utilization parameters. For completion of a particular job order, each machine is allocated with specific operations. Thus, the number of jobs assigned to each machine is dependent upon the machining time. The make span duration to complete the overall work of the job order is minimized by proper handling of busy machines [37]. To optimize the machining time, the nonlinear algorithm is used to solve the scheduling problems. The nonlinear problem is solved for the 'n' number of jobs which is allocated in the 'r' number of machines. This complex problem is solved by dividing the nonlinear problem into subproblems. Each subproblem is considered as finishing one particular job. A goal-oriented approach is adapted to solve this subproblem for minimum manufacturing time. Fig. 2 illustrates the flexible manufacturing system layout.

Another method such as autonomous scheduling is done by a knowledge-based approach in flexible manufacturing systems. This method eliminates the complex subroutine problem-solving methods. The database, knowledge base and interference engine are the main three steps [38] in the knowledge-based approach of autonomous scheduling. The database has information regarding the objective function of the goal. The knowledge base has the domain-based precise information of the system. The inference engine manages the knowledge base and database to execute the automatic job scheduling operations.

Artificial intelligence technology is another innovative method for job scheduling operations in manufacturing systems. Artificial based scheduling method [39] has the capability to learn from the previous scheduling operation and fine-tune the future job scheduling. The basic characteristic of this method, the scheduling operation is directly utilized from the previous similar work plan without solving all the complete job operations. The job scheduling operators learn new methods for every improved scheduling operation. The subroutine is also utilized in this method to enhance the scheduling

84

characteristics. The knowledge learnt from each scheduling operation is implemented in the next scheduling operation [40]. Thus, the learning scheduling operators are utilized in the organization by artificial intelligence techniques.

1.4 CNC machining parameters optimized by Artificial intelligence concepts

CNC machining factors must be optimized to obtain maximum productivity and better-finished product quality. An artificial intelligence technique is one of the methods to analyze and optimize the CNC machining behaviours [41]. For the characterization of machining factors, the real-time observation of several factors must be done. The sensors and transducer are adapted to monitor the real-time data observation of several CNC machines. The observed data is stored in the central computer through an online network [42]. In online mode, the machining factors are analyzed to find the best fit in the real boundary conditions based on the sensor data. The machining parameters such as acoustic noise, tool wear, the cutting force acting on the tool, the temperature of the tool are continuously monitored by the feature sensors. This absorbed data is the key element for predicting machining factors, tool conditions, and job surface finish. The sudden breakdown of machine and tool broken is predicted previously by artificial intelligence based on the gathered information from sensors to ensure the safety and robustness of the machining process within minimum simulation time [43].

1.5 Artificial intelligence for additive manufacturing

Artificial Intelligence is one of the advanced techniques handled by machines specifically with computer-based systems [44]. Artificial Intelligence aims to voice recognition, language handling, solving complicated problems and export systems. Artificial intelligence is made into several subdivisions, distributing a crucial method to resolving problems yet concerned in various areas, spanning expert methods, from machine learning. In this section, Artificial intelligence utilization in various stages of additive manufacturing is discussed. A component manufactured in an additive manufacturing machine is shown in Fig. 3.

85

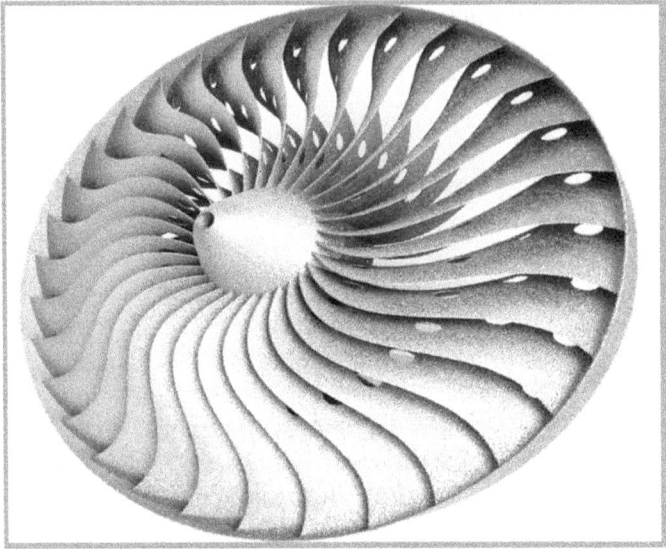

Figure 3. Additive manufacturing machine component

Printability is the ability to increase a 3D sample through 3D print-ers. Any three-dimensional article expects the 3D printer method to print. When compared to manufacturing processes, the develop-ment and application of 3D printers are minimal because of the ge-ometrical credit, time depletion and requirement of some specific materials. The 3D pattern can be made to minimize the complication of goods manufacturing in the best way [45]. The printability check-er is used to predict if an article is worthy to be a 3D printer. In a printability checker, a service-oriented problem is modelled as a feature extractor, a printer manager and a verifier engine.

To minimize the difficulty of manufacturing prototypes, the three-dimensional model must be created in an optimal way. The printa-bility checkers by learning techniques give clear information about the feasibility of the model to become the final product [46]. The decision is based on several factors the product characteristic done in the printability checker. The prediction of difficulties is based on the several indicators available in the trial run. The solver decodes the constraints in the verifier and transfers the data to the printer manager. Before going to the actual 3D printing the several factors, feasibility and difficulties are checked in these three components.

86

The implementation of learning techniques empowers the software tools to give the reliability, feasibility of products under several boundary conditions. The various available models for the other option to the complexity is provided by the system. Thus, the printable model is effectively optimized by the learning technique under the artificial intelligence method [47].

Several researchers proposed machine learning techniques for the autonomous rule adjustment; especially the factors opted to check the product printability. The support vector machine is used to train the computer tools for calculating the model valuation. A better solution is achieved by the classification of several parameters into a finite number of groups. Each group has similar models, methods and features. The assessment time is effectively minimized by these classification techniques available in machine learning systems. Several experimentation results proved the effectiveness of classification approaches in machine learning techniques. Thus, the negative parameters are reduced in the feature extraction time of 3D models.

In real-time production, the single indicator is not trustworthy to identify the complexity prediction. Multiple indicators must be included to find a suitable solution. The production time, raw material, price, shape and size of the geometry are the basic indicators that affect the overall output. The experimentation techniques are very costly to know the better result in every case. The simulation technique under intelligent tool gives worthy results at a low price. The provision for giving several ranges of inputs is possible in simulation tools. The advanced artificial intelligence learning techniques expressed the real-time solution under several ranges of factors.

1.6 Feature recognition in manufacturing system assisted by artificial intelligence

Feature recognition is the important parameter in the product design stage, analysis stage, process planning stage, model storage and retrieval stage. Industries expect highly efficient methods to develop the product in the product life cycle management. The industrial experts give proper guidance to develop the new product stage by stage in the product life cycle. The data must be shared among several experts from various fields for efficient product development.

Several field knowledge is required, mainly feedback from marketing people, machine handlers who machine the product, design engineers, quality control and so on. The computer-aided process planning activity is required to transfer the 3D model to the workflow chart for the industry lane. In design itself, the physical structure and ergonomics aspects are clearly handled. While giving input to the software tools, the automated feature recognition of the components is better for simple components. In the automated process planning schedule, two important conflicts occur [48].

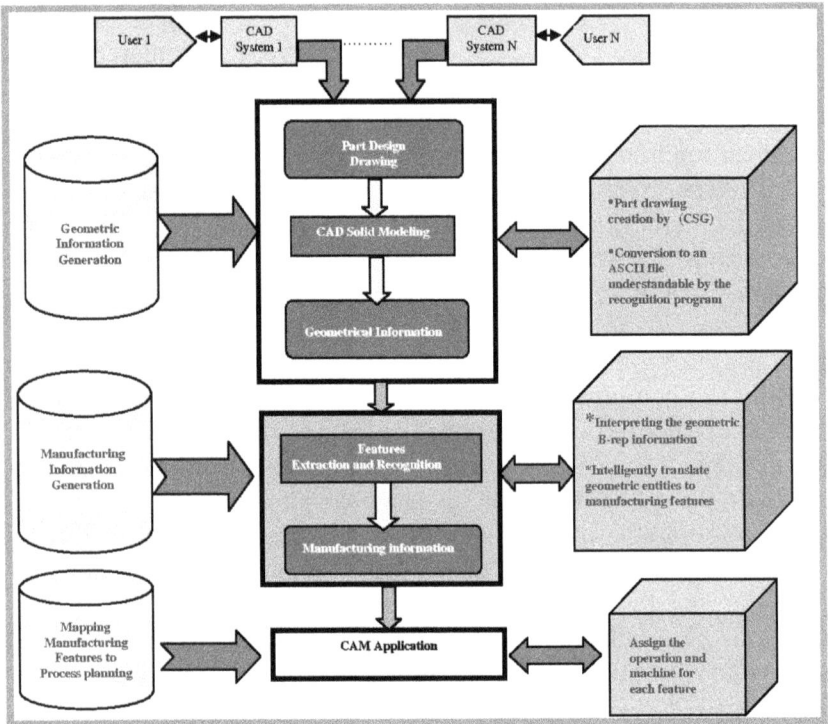

Figure 4. Feature recognition techniques

Fig. 4 shows feature recognition techniques flowchart. Mainly intersecting features and multiple interpretations of solid models in terms of features is the global two issues in automated process planning. Similar products are scheduled from the default standard model from the software tool library. The automated process planning schedule is difficult for complex geometry shapes. Artificial intelligence technology is the better way to utilize the learning methods for the process planning operations. The Knowledge-based Model Generator produces that model autonomously with the help of functionalities from the object model. It includes information directly available from the solid model and also factors evaluated by the object model (e.g. accessibility, concavities, etc.,). A planning model contains an initial state, a goal state and a set of action schemas for the current creating plans. Features are shown as action plans in the feature library, while the initial and goal states are produced by the Knowledge sourced Model Generator.

1.7 Digital twin technology implementation by artificial intelligence

The recent production system expects a higher rate of productivity and product variety. The demand must be fulfilled under the minimum delivery durations. The unexpected job orders or maintenance affect the delivery of the product on time. The frequent changes of the production schedule may adapt in the production lane due to the dynamic conditions. Rescheduling under human guidance based on real-time information is very difficult. And so the use of very high technological concepts is implemented to handle the production schedule in an autonomous manner under software guidance. The software tools must be highly intelligent for handling the variety of inputs under several boundary conditions. As a result, digital twin manufacturing technology is evolved for handling the rapid demand for the demand-based production system. The Internet of things, Big Data technologies, higher-level computation methods, Cyber-physical systems, central cloud computing deliver the foundations for Digital Twin and Artificial Intelligence applications that support the digital twin technologies concepts to real-time practices [49]. The real-world problems are simulated in the virtual world with no physical test case. The interconnection between the real and virtual parameters is properly given. This reduces the expenditure for designing the product or scheduling the process plan. The overall performance is predicted in the virtual simulation. Also, the digital twin technology enhances the automation of several process parameters. The designs, production plan, scheduling and cost estimation are automated with the assistance of artificial intelligence techniques. Several types of research correlated the experimental and virtual simulation results for better utilizing of cost less simulation practice in software tools. The simulations itself used to evaluate and verify the process model towards the machining process optimization. Artificial technology gives more real and affordable results in the dynamic decision process in the various business models. Machine learning techniques, a subset of artificial intelligence techniques accomplished for knowledge extraction for the real-time decision-making conditions. The structured and unstructured volume of information is used in this machine learning technique. Fig. 5 explains the digital twin technology concept.

Figure 5. Digital twin technology

The amount of data collected from the shop floor is huge. This collected data is given as input to the machine learning techniques to train the model. Human labelling is required in supervised learning techniques. The information on the normal and abnormal operating conditions is fed into the learning techniques. The gathered information has required data as well as very basic low-level information. The waste data is segregated but it is a time-consuming process in dynamic environments. Manual labelling is a time-consuming process. With the assistance of artificial intelligence techniques, the limitation of human intervention in supervised learning methods is eliminated. The time and cost is saved by implementing artificial intelligence techniques in manufacturing systems. By the virtual simulation methods, the proper data structure is obtained to train the machine learning models. Thus, machine learning is the cost-effective approach for complex data analysis and the time-effective way [50].

1.8 Conclusion

In this book chapter, the difficulty in the various traditional manufacturing systems and the effect of artificial intelligence techniques to handle those complexities are discussed. Especially in additive manufacturing techniques, the mass reduction, effective shape for the rigid body and minimum manufacturing time to create the part is highly influenced by artificial intelligence techniques. The real-time dynamic control over the machining operations based on the customer need is effectively adapted by this technique. This accelerated the manufacturing system for technological advancement by the industrial revolution 4.0. The knowledge-based approach in a flexible manufacturing system to handle the onsite changing job order conditions. The optimum route of process flow is identified by implementing artificial intelligence planning methods. Further, the knowledge-based approach blended with learning techniques make thrust worthy scheduling routes and minimized the major complexity in the flexible manufacturing system. The CNC machining parameter optimization by artificial intelligence is another milestone in the manufacturing system. The breakdown of machine and tool worn-out is predicted by the surrounding environmental variables monitoring. The sensor is fitted in each machine to observe environmental factors such as acoustic vibrations, temperature and tool force. These parameters are fed into the artificial intelligence tools for further analysis. This information is effectively utilized for better machining characteristics of CNC machines to attain higher production without compromising the quality of the products.

References

[1] Appadurai, M., & Velmurugan, V. (2015). Performance analysis of fin type solar still integrated with fin type mini solar pond. *Sustainable Energy Technologies and Assessments*, 9, 30-36.
[2] Appadurai, M., Prasath, A. B., Thanu, M. C., & Richard, S. (2015). Determination of dual phase forced convective heat transfer of nano fluids by means of CFD. *Int. J. Appl. Eng. Res*, 10(12), 32627-32635.

[3] Appadurai, M., & Velmurugan, V. (2017). Experimental analysis of stepped basin pyramid solar still integrated with mini solar pond. *Desalination and Water Treatment*, 84, 1-7.

[4] Appadurai, M. Studies on enhancing the performance of modified solar still with forced mode of water circulation and with additional heat input.

[5] Mangalaraj, A., &Vellaipandian, V. (2018). Performance analysis of modified solar still with forced water circulation. *Thermal Science*, 22(6 Part B), 2955-2964.

[6] Appadurai, M., Baburaj Prasath, A., Starwin Jedidiah, S., Daniel Lawrence, I. (2015). Experimental Analysis of Parabolic Disc Type Solar Still. *Int. J. Appl. Eng. Res*, 10(51), 688-690.

[7] Appadurai, M., & Raj, E. F. I. (2021, February). Finite Element Analysis of Composite Wind Turbine Blades. In *2021 7th International Conference on Electrical Energy Systems (ICEES)* (pp. 585-589). IEEE.

[8] Appadurai, M., Raj, E. F. I., &Venkadeshwaran, K. (2021). Finite element design and thermal analysis of an induction motor used for a hydraulic pumping system. *Materials Today: Proceedings*, 45, 7100-7106.

[9] Fantin Irudaya Raj, E., & Appadurai, M. (2021). Minimization of Torque Ripple and Incremental of Power Factor in Switched Reluctance Motor Drive. In *Recent Trends in Communication and Intelligent Systems: Proceedings of ICRTCIS 2020* (pp. 125-133). Springer Singapore.

[10] Raj, E. F. I., & Kamaraj, V. (2013, March). Neural network based control for switched reluctance motor drive. In *2013 IEEE international conference on emerging trends in computing, communication and nanotechnology* (ICECCN) (pp. 678-682). IEEE.

[11] Sijini, A. C., Fantin, E., & Ranjit, L. P. (2016). Switched Reluctance Motor for Hybrid Electric Vehicle. *Middle-East Journal of Scientific Research*, 24(3), 734-739.

[12] Chouhan, A. S., Purohit, N., Annaiah, H., Saravanan, D., Raj, E. F. I., & David, D. S. (2021). A Real-Time Gesture Based Image Classification System with FPGAand Convolutional Neural Network. *International Journal of Modern Agriculture*, 10(2), 2565-2576.

[13] Raj, E. F. I. (2016). Available Transfer Capability (ATC) under Deregulated Environment. *Journal of Power Electronics & Power Systems*, 6(2), 85-88.

[14] Deivakani, M., Kumar, S. S., Kumar, N. U., Raj, E. F. I., & Ramakrishna, V. (2021, March). VLSI Implementation of Discrete Co-

sine Transform Approximation Recursive Algorithm. *In Journal of Physics: Conference Series (Vol. 1817, No. 1, p. 012017).* IOP Publishing.

[15] Jenish, I. (2021). I. Jenish, M. Appadurai, E. Fantin Irudaya Raj" CFD Analysis of Modified Rushton Turbine Impeller. *International Journal of Science and Management Studies* (IJSMS), 4, 13.

[16] Raj, E. F. I., & Appadurai, M. (2021). The Hybrid Electric Vehicle (HEV)—An Overview. Emerging Solutions for e-Mobility and Smart Grids, 25-36.

[17] Shafali Jain, BalarengaduraiChinnaiah, M P Rajakumar, E.Fantin Irudaya Raj, & Gujar AnantkumarJotiram (2020). Improved IoT-Based Control System Combined with an Advanced Control Management Server-Based System *Journal of Green Engineering*, 10(10), 8488-8496

[18] Priyadarsini, K., Raj, E. F. I., Begum, A. Y., &Shanmugasundaram, V. (2020). Comparing DevOps procedures from the context of a systems engineer. *Materials Today: Proceedings.*

[19] R. Mathumitha, V. Suvitha, & E. Fantin Irudaya Raj (2020). Design of Controllers for Buck Converter *Journal of Interdisciplinary Cycle Research*, 12 (III), 104-115.

[20] N. Deivanayagam, M. Sathishkumar, & E. Fantin Irudaya Raj (2020). Comparative Review of Fuzzy Based Regenerative Braking in Electrical Vehicle. *Journal of Interdisciplinary Cycle Research,* 12 (III), 95-103.

[21] E. Fantin Irudaya Raj (2018), Review of Different Types of Batteries Used in Electric Vehicles and Their Charging Methods. *International Journal of Current Engineering and Scientific Research*, 5 (12), 110-113.

[22] E. Fantin Irudaya Raj (2015), Hybrid Power System Model Using MPPT Algorithm for Wind and Solar System. *International Journal of Innovative Research in Technology*, 1 (12), 1606-1614.

[23] Chien, C. F., Dauzère-Pérès, S., Huh, W. T., Jang, Y. J., & Morrison, J. R. (2020). Artificial intelligence in manufacturing and logistics systems: algorithms, applications, and case studies.

[24] Stalin David, D., Nageswara Rao, G., Swain, M. P., Venkatesh, U. S., Fantin Irudaya Raj, E., & Saravanan, D. (2021). Inflammatory syndrome experiments related with COVID-19. *Turkish Journal of Physiotherapy and Rehabilitation*, 765-768.

[25] Krishna, B., Amuthavalli, G., StalinDavid, D., Raj, E. F., & Saravanan, D. (2021). Certain Investigation of SARS-COVID-2-

Induced Kawasaki-Like Disease in Indian Youngsters. *Annals of the Romanian Society for Cell Biology*, 1167-1182.

[26] Ch, G., Jana, S., Majji, S., Kuncha, P., &Tigadi, A. (2021). Diagnosis of COVID-19 using 3D CT scans and vaccination for COVID-19. *World Journal of Engineering*.

[27] Agarwal, P., Ch, M. A., Kharate, D. S., Raj, E. F. I., &Balamuralitharan, S. (2021). Parameter Estimation of COVID-19 Second Wave BHRP Transmission Model by Using Principle Component Analysis. *Annals of the Romanian Society for Cell Biology*, 446-457.

[28] Bullers, W. I., Nof, S. Y., &Whinston, A. B. (1980). Artificial intelligence in manufacturing planning and control. *AIIE transactions*, 12(4), 351-363.

[29] Cadavid, J. P. U., Lamouri, S., Grabot, B., Pellerin, R., & Fortin, A. (2020). Machine learning applied in production planning and control: a state-of-the-art in the era of industry 4.0. *Journal of Intelligent Manufacturing*, 1-28.

[30] Jain, D., Sangale, D. M. D., & Raj, E. (2020). A Pilot Survey Of Machine Learning Techniques In Smart Grid Operations Of Power Systems. *European Journal of Molecular & Clinical Medicine*, 7(7), 203-210.

[31] Anbuchezhian, N., Devarajan, B., Priya, A. K., &Rajeshkumar, L. (2020). Machine Learning Frameworks for Additive Manufacturing–A Review. *Solid State Technology*, 63(6), 12310-12319.

[32] Raj, E. F. I., & Balaji, M. (2021). Analysis and classification of faults in switched reluctance motors using deep learning neural networks. *Arabian Journal for Science and Engineering*, 46(2), 1313-1332.

[33] Gampala, V., Kumar, M. S., Sushama, C., & Raj, E. F. I. (2020). Deep learning based image processing approaches for image deblurring. *Materials Today: Proceedings*.

[34] Levi, P. (1987, March). Principles of planning and control concepts for autonomous mobile robots. In *Proceedings. 1987 IEEE International Conference on Robotics and Automation* (Vol. 4, pp. 874-881). IEEE.

[35] Antsaklis, P. J., &Passino, K. M. (1993). Modeling and analysis of artificially intelligent planning systems. *An Introduction to Intelligent and Autonomous Control*, 191-214.

[36] Jain, P. K., & Mosier, C. T. (1992). Artificial intelligence in flexible manufacturing systems. *International Journal of Computer Integrated Manufacturing*, 5(6), 378-384.

[37] Bakakeu, J., Tolksdorf, S., Bauer, J., Klos, H. H., Peschke, J., Fehrle, A., ... & Franke, J. (2018). An artificial intelligence approach for online optimization of flexible manufacturing systems. In Applied Mechanics and Materials (Vol. 882, pp. 96-108). *Trans Tech Publications Ltd.*

[38] Spano Sr, M. R., O'Grady, P. J., & Young, R. E. (1993). The design of flexible manufacturing systems. *Computers in Industry*, 21(2), 185-198.

[39] Young, R. E., & Rossi, M. A. (1988). Toward knowledge-based control of flexible manufacturing systems. *IIE transactions*, 20(1), 36-43.

[40] Kaula, R. (1998). A modular approach toward flexible manufacturing. *Integrated Manufacturing Systems.*

[41] Park, K. S., & Kim, S. H. (1998). Artificial intelligence approaches to determination of CNC machining parameters in manufacturing: a review. *Artificial Intelligence in engineering*, 12(1-2), 127-134.

[42] Ramli, K. D. N. R. (2014). Application of artificial intelligence methods of tool path optimization in CNC machines: A review. *Research Journal of Applied Sciences, Engineering and Technology*, 8(6), 746-754.

[43] Abellan-Nebot, J. V., &Subirón, F. R. (2010). A review of machining monitoring systems based on artificial intelligence process models. *The International Journal of Advanced Manufacturing Technology*, 47(1), 237-257.

[44] Heiden, B., Alieksieiev, V., Volk, M., &Tonino-Heiden, B. (2021). Framing Artificial Intelligence (AI) Additive Manufacturing (AM). *Procedia Computer Science*, 186, 387-394.

[45] Paraskevoudis, K., Karayannis, P., &Koumoulos, E. P. (2020). Real-time 3D printing remote defect detection (stringing) with computer vision and artificial intelligence. *Processes*, 8(11), 1464.

[46] Siemasz, R., Tomczuk, K., &Malecha, Z. (2020). 3D printed robotic arm with elements of artificial intelligence. *Procedia Computer Science*, 176, 3741-3750.

[47] Kaleem, M. A., & Khan, M. (2020, January). Significance of Additive Manufacturingfor Industry 4.0 With Introduction of Artificial Intelligence in Additive Manufacturing Regimes. In 2020 17th *International Bhurban Conference on Applied Sciences and Technology (IBCAST) (pp. 152-156). IEEE.*

[48] Marchetta, M. G., &Forradellas, R. Q. (2010). An artificial intelligence planning approach to manufacturing feature recognition. *Computer-Aided Design*, 42(3), 248-256.

[49] Zhou, G., Zhang, C., Li, Z., Ding, K., & Wang, C. (2020). Knowledge-driven digital twin manufacturing cell towards intelligent manufacturing. *International Journal of Production Research*, 58(4), 1034-1051.

[50] Kaur, M. J., Mishra, V. P., & Maheshwari, P. (2020). The convergence of digital twin, IoT, and machine learning: transforming data into action. In *Digital twin technologies and smart cities* (pp. 3-17). Springer, Cham.

Chapter 7

Review-Real Time Smart Energy Meter and Load Automation Using IOT

[1]Ajesh F, [2]Aswathy S U, [3]Felix M Philip, [4]Shermin Shamsudheen,
[1]Department of Computer Science and Engineering, Sree Buddha College of Engineering, Alappuzha, Kerala, India
[2]Department of Computer Science and Engineering, Jyothi Engineering College, Thrissur, India
[3]Department of Computer Science and Information Technology, Jain (Deemed-to-be University), Kochi, Kerala, India
[4]Faculty of Computer Science & Information Technology, Jazan University, Saudi Arabia

Abstract: The power utilization in family is expanding quickly through time because of numerous issues. Since clients don't know about their power utilization information and tax continuously power is squandered. Which lead them to pay high measure of cash and because of the high duties of power a few purchasers are compelled to do power burglary. Because of this the use of shrewd electric meters is essential to track and record the ongoing power utilization of a family unit. Utilization of Internet of things (IOT) will make the assortment, transmission and examination of electric utilization information among clients and utility's quick and simple cycle. This paper centers around introducing the various highlights and advancements that are being utilized on flow IoT based keen power meters and dependent on the current plans. We have proposed minimal effort and energy effective brilliant energy meter plan with remote correspondence. In view of the new proposed framework we have set up remote based shrewd metering framework were clients can without much of a stretch their ongoing utilization.

Introduction

Starting at 30 November 2017, the power market in India had one public framework with an introduced limit of 330.86 GW. Inexhaustible force plants represented 31.7 percent of the general introduced limit. The gross power created by utilities in India was 1,236.39 T Wh during the long term, and the general power creation on the planet was 1,433.4 T Wh. During the years 2016-2017, the gross energy consumption per capita was 1,122 kWh [1-7]. As the world's third-largest producer of electrical energy, India ranks fourth in the world in terms of control utilisation. In 2015-16, it was estimated that the agrarian region consumed 17.89 % power consumed by all countries. Regardless of India's lower energy costs, power utilization per capital is restricted comparative with numerous nations. India's capacity age limit is excess, however there is an absence of a satisfactory foundation for providing power to every destitute person [11-13]. The Government of India has dispatched a plan called "Power for All" to build up the foundation to flexibly satisfactory power to all the destitute individuals in the nation by March 2019. By improving the essential framework, this plan will guarantee a ceaseless and continuous flexibly of capacity to all ventures, family units, and business foundations. The Government of India has a joint obligation with states to share financing and produce by and large monetary development [4].

Petroleum products, particularly coal, which created around 66% of all power in 2016, overwhelm the power area in India [8-9]. Notwithstanding, the public authority is simply expanding its dedication in clean energies. By 2027, with the charging of 50,025 MW coal-Based force plants being chipped away at and the accomplishment of 275,000 MW joined introduced practical force limit, the draft National Electricity Plan 2016 planned by the Government of India sees that the nation needn't waste time with extra non-unending power plants in the utility market [10-15]. With 4.8% of the worldwide offer, India has become the world's third biggest power maker. Inexhaustible power represented about 28.43% of the general power delivered and non-environmentally friendly power represented about 71.57% [1]. The basic essential for having an agreeable existence is power. It is expected to be utilized and taken care of appropriately. At the present time, the Electricity Board human executive visits the inhabitants to take the readings from the energy meter and really make the bill during the stream month. There is a disturbing expansion in the interest for power. The treatment of power upkeep and requests is subsequently getting progressively confounded [16]. Thus, it is important to spare however much power as could be expected right away. Energy spared is created by energy, and in this way power sparing move ought to be made. There are strong frameworks, for example, energy the executives and house automation to spare energy utilization [19]. The critical targets for a superior framework are checking, streamlined power use and force squander decrease. The arrangement of energy the board diminishes the general expense. These investment funds could be brought about by improved utilization of human force, energy investment funds just as framework disappointment. With the assistance of advance innovations and applying in this true issue brings out a lot of effective arrangements that can be ease and proficient for buyers [17-18].

2. LITERATURE REVIEW

2.1 Monitoring and monitoring of loads based on IOT Smart Energy Meter Billing

In this article, we have ceaselessly tended to the proposition and improved a decent perception and control framework for energy meters. The Remote Meter Reading Device [22] is made to screen power, along these lines decreasing expense of creation. With a prepaid allotment framework, separation energy meters were organized. The structure that utilizes a web-worker, with web of things, to plan virtual instrument programming [34]. When all is said in done, the System investigates the imperativeness and utilization of solidarity models. What's more, a prepared SMS by means of GSM is normally shipped off a versatile master concerned on the off chance that the conditions are not plain also, and customers can then quickly pay for the following month toward the month's end, with current utilization figures [20-21]. The system is likewise ready to see status and move subtleties to a web worker. The master concerned will screen and build up the framework by utilizing the powerful contraptions web-based interface. The page we are going to use is enigma express guaranteed by adding closure of username and secret key by guaranteed API keys. In areas where genuine closeness is foolish all an ability to manipulate the gadgets[23], this framework finds a wide application. The arrangement would work with the ARM processor used for the use of the sensor module and other requirements for correspondence. The framework provides an absolute, immateri-

99

al effort, earth shattering and easy-to-use strategy for propelling appliance seeing and managing.

A repetitive loop is the current model and it takes a lot of development. The system proposed takes out the use of work and it is a cycle that is savvy and effective. The proposed structure gives the details about the energy use on regular timetable, charging and portion via IoT, pre-recommendation of shut down nuances, prepared systems when the energy use outperforms past very far and the withdrawal of power through a message when the private are out of station to prevent the wastage of energy. This framework chiefly screens electrical boundaries of apparatuses and hence computes the units devoured. As WSN's are having numerous points of interest, here we have planned keen meters foreseeing the utilization of intensity utilization. In any case, it is minimal effort, adaptable, and strong framework to persistently screen and control dependent on customer necessities, Wi-Fi innovation for systems administration and correspondence, since it has less-force attributes, which empower it to be generally utilized in home and building conditions Figure 2.1 [24-25].
.

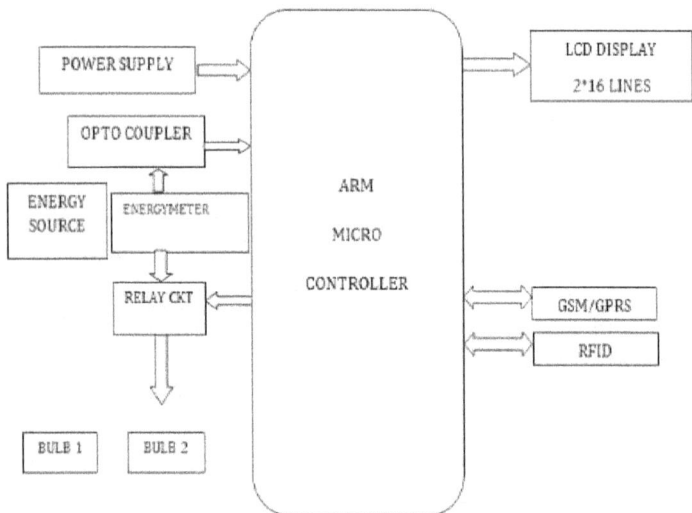

Fig. 2.1 block framework of proposed system

2.1.1 GSM Enabled Smart Energy Meter and Automation of Home Appliances

In this research, a brilliant energy system for private customers is developed, as well as a sharp switch board that can reduce the need to ascend to clever devices in order to make the structure more financially viable. An additional development was the creation of a virtual instrumentation that can be used as an In-Home Display (IHD) for the Energy Management System (EMS) on any PC. The term "Smart Energy Meter" refers to a technology that monitors the consumption of energy over specific specified periods of time and transmits the information to the inspection utility as well as to the bosses and for charging purposes [26].Not at all like the ordinary end utilizes a pivoting wheel for estimating energy savvy meter conveys sensors for the measure current and force quality. Dynamic force, voltage marker for programmed stumbling of burden showed in the IHD. Meter readings are sending X-Bee and client's cell phone and client. Consequently, the Smart Energy Meter sets up correspondence between the client and u useful highlights of the Smart Energy Meter [24-28].

•Voltage reading, current reading and m distinction between the current and volt
•Transmitting the data to client's cell phone utilizing GSM.
•Transmitting information to clients IHD utilized Remote control of home apparatuses us cell phone.

The Hall Off the proposed model in the proposed clever meter, is used deftly as opposed to the present transformer to a synchronized voltage. These are then discussed by measuring the zero convergence of current and voltage as a basic dedication to the power factor is evaluated. Small PC controllers[29] are used for energy consumption. For the p PC to fill in as the IHD t use of committed things, the agreed energy small scale controller similar to PC for plotting energy use mutt Instrumentation was made. It provides energy usage during the unmistakable use of GSM for co-customers and utilities in the Smart Energy Meters. The architecture of the device under consideration is depicted in Fig.2.2.

By then taking the best possible action of the data found in the SMS, the data is sent to the utility laborer for evaluation response as short

(SMS) from the utility specialist via the splendid data metre. This deftly assistants in a client's control when the bill shortens manual intercession. The regular month-to-month bill and the piles of the splendid energy metre will be updated. In addition, the stacks are switch energy usage exceeds the set philtre customer provided the option of customized trading of mechanical weight assemblies or when the customer is forewarned of incredible use[27-28]. In addition, the consumer is far from exchanging machines with GSM mobile. The proposed model can be used to test the power consumption of a nuclear family by voltage, current and figure 2.2.Theose data can be further used for processing and analyzing for several uses.

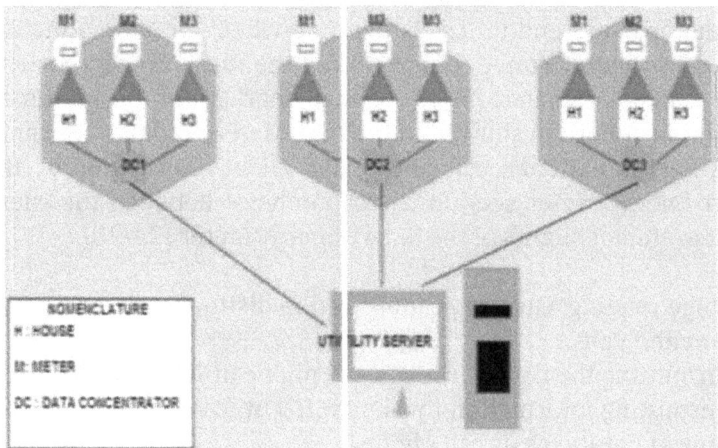

Fig 2.2 Smart energy meter

2.1.2   Smart Energy Meter Tracking and Theft Detection based on IoT using ATMEGA

The following are the primary destinations of this system, as depicted in detail: loss of control generates the prices imposed to consumers and has the potential to provide actual well-being outcomes[32]. Traditionally, real tests of changing apparent seals by field workers and the use of balance metres have been used to identify power burglaries. Recognize the breakdown by giving the holder a ready SMS. Send metre readings to the owner and rate each month. A proficient Internet of Things (IoT) is defined in addition to these arrangements, which images the clients'

102

worldwide association environment and enables them to display the status of metre perusing and burglary affiliations uniformly from wherever whenever. In terms of cost and security, theft of power greatly affects customers.we consider that providers are not adequately urged by the existing administrative framework to be vigilant in separating burglaries[39]. In this paper, we refer to the proposed new flexibly permit commitments to improve burglary handling courses of action and the proposed portion of Distribution Network Operators (DNOs) in the handling of robbery when suppliers are not obliged to do so[30]. We also advise on steps and guidelines for additional approaches to assist providers in the discovery, detection and prevention of theft[34-38]. For all the system, it is beneficial to deter burglaries and the medium associated around the world to properly image the metre perusing to its customers. Power breakdown generates the costs charged by consumers which can result in real well-being. This allows suppliers to misallocate costs that can twist rivalry and impede the proficient activity of the industry.

The expenses looked at by a power supplier in its customers' recognition of power burglary could be more prevalent than the total business expenses. Power theft is one such instance. When a client complains power theft, the service provider may be liable for identifying the age, organisation, and compensation costs connected with bringing assessments of the quantity of power taken by that customer to a settlement agreement. Then again, this operation does not promote an escalation of costs at the level of the organization in general. Distinguishing power theft has usually been tended to by real checks by field personnel for modifying apparent seals and by using balance metres. They are missing, despite the fact that these techniques decrease unmeasured and unbilled use of power. To be sure, it is possible to effectively circumvent altering apparent seals, and despite the fact that equilibrium metres can realize that a few clients are deceitful, they cannot precisely identify the perpetrators.The higher-goal data collected by them is seen as a promising advancement that will complement traditional recognition methods, regardless of the security vulnerabilities of shrewd metres. Metering, billing and assortment steps and the discovery of extortion and unmetered associations can be improved[33-34]. Basic burglary strategies vary from trading off metre protection to legally interfacing burdens with power dissemination lines. Due to problematic levels of inspection and specifications, default of instalments was a major issue. The lack of creativity and the lack of motivating forces for merchants were the major supporters of this issue. To calculate the approaching current from the power metre, the CT (Current Transformer) sensor is used and appears on the LCD monitor. If you add burden to the Power Meter, it burns-with some force, this value appears through sequential correspondence on the LCD just like the PC.Voltage Sensor is used to flexibly discover the voltage level from simple and appeared in LCD. This voltage calculation is shipped using a TTL-USB converter from the PC. The owner sends out this warning mail as soon as possible. The message contains the most recent estimations of current, voltage, and consumption. The units are represented by the four parts of digit 7.

Driven Monitor, Reset the device to 0.0 once the metre is taken If an unapproved person uses control, it provides the proprietor with a momentary SMS warning. In the Internet of Things (IoT) model, in a certain structure, a large amount of the living and non-living things that encompass us would be on the internet. Driven by the popularity of contraptions attracted by wireless mechanical turn of events, such as[15], Wireless Bluetooth, Radio Frequency Recognition, Wireless-Fidelity, embedded sensor, IoT has stepped out from its beginning stage and is truly on the verge of transforming the present fixed between the internet into an internet connected inside and out. There are practically nine billion related gadgets by and by, and it is estimated that about fifty billion gadgets will be directly linked by 2020[37]. The planet faces such an environment today that presents challenges.The standard issue discussed by our general public is the energy crisis. One of the solutions to this problem is an effective structure to regulate and track power use. The elimination of the use of force in families is one procedure by which the current energy crisis can be resolved. Buyers are increasing increasingly and there is a distinct expansion of inconvenience in power offering divisions. By offering them an ideal game plan, buyers must be empowered: - For example, the probability of IoT (Internet of Things) metres and then again master centre end can be taught similarly about force robberies using thievery area unit and PLC modem[18-19]. The possibility of IoT metres including four in-

104

teresting units thrived by holding above segments: Micro-controller unit, Theft ID unit, Meter Analysis and correspondence unit. ATMEGA328P Micro-controller-based arrangement and energy metre execution using IoT and theft control thinking is portrayed in the paper.by supplying equipment, the customer can track energy usage in units from a tab. The energy meter-related thievery area unit will say the association side when metre alteration takes place in the energy metre and will submit theft recognising information through capable applications and perceived burglary will appear on the terminal window at the end of the pro core. Indeed the current demand requires indirect access to the contraption characteristics in a secure manner. Figure 2.3 One of the potential ways to deal with the effort is to connect a contraction (energy metre) to the web by supplying it with efficiency.

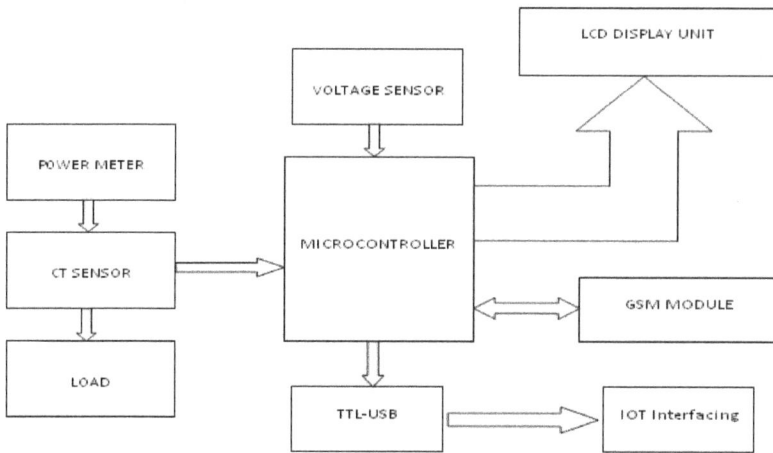

Fig 2.3 block diagram

2.2 SMART WIRELESS METERS OF ELECTRONIC ENERGY

This paper proposes that a sharp, consolidated notice system for power use has been introduced using open standard growth, enterprise and family stuff that screens voltage and current extent in an inaccessible structure viably. Here is a canny energy metering structure based on GSM using IOT that will supersede the ordinary way of having metres. Without the person visiting each building, they can accurately screen the metre readings. ATMEGA328-based force utilization control system that senses limits and displays on an LCD monitor. The meter readings are consequently sent on Cloud produced utilizing IOT. This framework will forestall the illicit utilization of power. It will give unadulterated straightforwardness in the framework. The primary objective of this framework is to lessen miss correspondence between the client and wholesaler. This framework will likewise assist with getting straightforwardness power bills [35]. It is more effective and can be actualized in minimal effort. Electrical metering instrument advancement has gained some astonishing ground from what it was

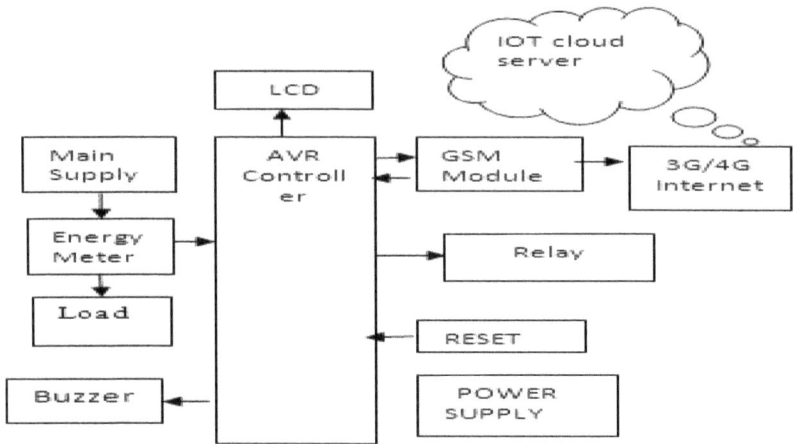

Fig. 2.4 Block diagram of Smart Energy Meter System

more than 100 years back. From the principal lumbering meters with heavy magnets and twists there have been various advancements that have achieved size and weight decline despite progress in features and subtleties. So, it is new idea in universe of Electricity estimation [27]. This idea isn't just valuable for power estimation yet in addition has the capacity to forestall abuse of power. Currently, owing to the colossal disparity in energy consumption and energy generation, the use of energy and circulation has become a major topic for discussion. At present the majority of the houses in INDIA has the customary mechanical watt hour meter and the charging framework isn't robotized. So, another framework was examined known as keen energy electronic meter which uses cloud in it. Cloud correspondence alludes to innovation that empowers machines to be organized so information can be uninhibitedly traded among these resources. It is a type of information correspondence that includes at least one substance that don't really need human connection or mediation during the time spent correspondence. It tends to be in two different ways one of them is Uplink to gather item and utilization data and another is Downlink to send directions or programming refreshes, or to distantly screen gear. In our framework, we are leveraging Arduino to determine energy consumption. The proposed energy observation framework has proved its ability to accurately calculate energy metre readings. The overall concept of the project was developed using an exceedingly rigorous technique, with an emphasis on minimising energy consumption factors [19]. This framework will ensure that communication between supplier and purchaser is transparent. The IoT-based energy metre was developed for determining how much power was consumed and how it displayed on the LCD. The spent energy is sent to the virtual terminal generated in PROTEUS via consecutive communication. This endeavour may then educate the board about idle and needless outings, accounting, and charging on the grounds that it provides accurate bookkeeping of units driven to avoid neglect. As shown in Figure 2.4.

2.3 Digital Smart Three Phase Energy Meter Based on Hall Effect Sensor

Currently, conventional electromechanical energy metres for electricity are being phased out in favour of electronic metres, just as they are in commercial applications. A hard and fast electronic three phase four wire energy metre is presented in the proposed device. In the automated space, all power assessments are taken. These readings are transmitted by strategy for distant GSM improvement to the convenience of the consumer. Updates of the electric power consumption data may be accessible to the customer on his compact[40]. The controller is used to control all metre elements. The proposed power sorting estimate clarifies hardware needs as the power factor is resolved in number, which reduces the need for the zero-crossing point marker circuit. Electrical Energy is getting important in human life. Since humans continued with presence and development absolutely depends on it, we never thought the presence without electrical power. As of now the energy management of business customers is becoming fundamental every day.

The primary objective of this system in the field of energy management (EMS) is to increase the efficiency of the electric power system. Convincing control of the structure necessitates a plethora of data regarding the constraints in a surprising number of main interests. This necessitates relocating the system's limitations to more convenient locations, such as power plants and substations. Voltage, stream, repeat, dynamic power, open power, power factor, dynamic energy, and responsive energy are the three-stage electric power system's critical constraints. The purpose behind this paper is to gather a KWH meter that can insightful the customers with messages. An energy metre or kilowatt-hour metre is a device that measures the amount of electrical energy consumed or generated by a dwelling, company, or equipment. Force is an ideal and appropriate approach to deal with energy transfer. When these metres are used in retail electricity, the utilities record the characteristics measured by them in order to create a voice for the force. Additionally, they may record numerous elements consolidating when the force was applied. In all domains, including power movement and charging, highly automated and dependable technologies are employed. The advancement of computerization is included into already operating advanced energy metres [22-26]. The proposed metre monitors voltage, current, active and reactive power, power factor, and power consumption for a three-stage load connected to the metre. This system utilises RTC to maintain the timetable and clock. Data relating to control use and billing information will be supplied to the customer's flexible device upon his request via the GSM modem. The ARM LPC2138 is the meter's brain. The construction is divided into three segments: the measurement unit, the controller, and the display unit. Sensors for the Entryway Effect Gauges Current and voltage. Yield-dependent 3Vpp bipolar sine waves from a Hessen chemist. To eliminate excessive repetition impedance, these indicators are utilised in an RC low pass technique with a cutoff repeat of 1 Khz. Additionally, these signals are sent to the ADC of the ARM controller, which detects only unipolar signals between 0 and 3.3V [42]. Thus, these bipolar sine waves are first converted to unipolar utilising an Op-Amp level shifter block (LM358). After correct sign enrichment, the inputs are routed to the ARM controller's integrated ADC. This ADC will convert these analogue signals to digital and display them on an LCD. The ARM's inbuilt RTC is utilised as the clock and schedule. Data about the customer's power consumption can be transmitted through a GSM modem connected to the ARM controller. The ARM controller features two

108

UARTs, one for ISP programming and another for GSM module interface. The power necessary for all sensors and sign trim squares is provided by a smoothly distributed power supply. The cutting-edge cycle includes the presentation of an ADC, a real-time clock, a timer, a UART, a GSM module, and an LCD. The cycle has initiated after which the establishment cycle tally would conclude. On the LCD display, voltage, current, power, power factor, and energy measurements are displayed. Persistent clock data (clock and time-table) appeared in a similar fashion [12]. If a GSM block occurs (when the controller approaches the SIM in the GSM module), the controller conducts an interrupt routine to send SMS to the enrolled adjustable number of customers to regulate use and billing. The sign trim square is denoted by the SCB. This document illustrates the configuration and operation of a three-phase energy metre. The Hall Effect sensor used for voltage and current discrimination exhibits linear yield ascribes. SMS messages describing control use and charging have been successfully sent from a customer's compact to a GSM module via a call to the module [11].

2.4 Sustained Energy Savings achieved by online smart metre analysis: findings from five communities

Together with consistently gas adroit metre data, hourly electric sharp metre data gives the data necessary to play out a savvy dis mixture of private energy usage. Elite, High Energy Audits (HEA) cloud-based inspection programming usually disengages energy usage time into seven groups and delivers this data to property holders through a 15-minute online survey. This degree of evaluation provides contract holders with ample knowledge to make remarkable, marginal adjustments in effort to achieve decreased energy usage in 66 percent of sharing households. Beginning in April 2011, this technique was used in five towns in the San Francisco Bay Area and finished in April 2012, with more than 200 families opting to participate consistently in the free programme. As distinguished from energy use in the previous year a hard and quick keep asset of 105,293 kWh was seen, or a typical 634 kWh per household for all individuals completing the audit. Most energy purchasers haven't the faintest idea how energy is truly used in their homes (Darby 2006). Most energy consumers don't have the faintest understanding about how their homes actually use energy (Darby 2006). They get and deal with a month-to-month tab that rotates equally, but has continued to expand over the long haul (AEP 2012)[9] anyway. In addition, the way they can reduce their energy consumption is unique to them as the fundamental data they have to catch up on is a single number at the end of the month. Affiliations like utilities, non-advantage green affiliations, local and other government affiliations are advanced by summarized proposals on increasing or cutting down indoor controllers and overriding sparkling lighting, however these summarized proposals do not have any kind of impact relative to all residences, are not upgraded, and discuss evolving levels of effort to complete. How do energy consumers understand what steps are usually appropriate for their situation? What energy consumption groups discuss their most noteworthy open gold reserve entryways? Where might it be a clever thought for them to focus their businesses? The arrangement rule behind the advancement of High Energy Audits (HEA) programming[19] was to emphasize the provision of significant details on certified energy usage.

Without the advantageous blending of government technology and growth that occurred in 2009, this improvement would not have been necessary. In 2006, the California Legislature enacted and Governor Schwarzenegger signed Assembly Bill 32, the Global Warming Solutions Act of 2006, establishing a 2020 target for ozone depleting substance (GHG) spreads decrease (ARB 2012).Therefore, culture in California has started to examine ways to cope with order fulfilment. In order to meet their GHG decline goals, companies with virtually no market success must concentrate on reducing private energy use. Analysis and Early Studies [7-9]. Steve Schmidt and Peter Evans experienced some time using contraptions, such as the Blue Line Whole House Monitor, HOBO loggers, and Kill-a-Watt metres, beginning in 2008, to assess energy usage in their homes and the homes of partners and neighbours. Two people from the Los Altos Hills Environmental Committee.Taking into account the criteria of AB 32, the Environmental Committee explored ways of coping with the decline in energy usage at home. Via their assessment and discussions with community residents, it had become evident that home renovations were not drawing on an after-effect of the enormous expense at this point as well and even impressively more profoundly, given the way that gigantic quantities of high-energy homes were new or late upgraded to Title 24 requirements and right now met high environs.

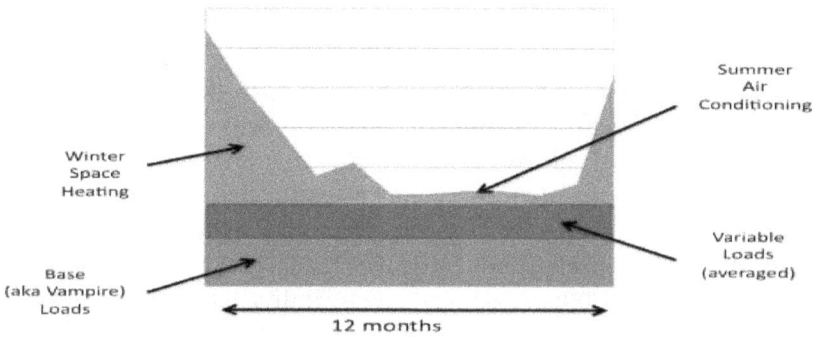

Fig 2.5 Energy Profile Showing Total Energy Use

2.5 Development and implementation of an Advanced Smart Energy Meter based on the Internet of Things (IOT)

The energy metre is an extremely important method for measuring power in the local, current, etc. atmosphere. To record the outright power use and subsequently for charge tally, the correct and valid evaluation of power with no mix-up is enormous. Considering this an innovative splendid energy metre is suggested in this paper plan and execution. The suggested shrewd energy metre relies on applications from the Internet of Things (IoT). The paper closely depicts its arrangement through its functioning. The Internet of Things (IoT) is the combination of actual computers, cars, home machines, and other things that are interconnected with equipment, programming, sensors, actuators, and networks of associations that assemble and share data with these things.Through its embedded figuring framework, anything is particularly conspicuous and can communicate within the current Internet establishment anyway. In spite of reduced human involvement, the IoT licenses objects to be recognised or regulated indirectly through established association institutions, allowing open entrances for clearer coordination of the real world into PC-based systems, and achieving enhanced profitability, accuracy and favorable budgetary role. The development transforms into an example of the more comprehensive class of computerized real structures when IoT is extended with sensors and actuators, which similarly includes headways, for example, swift cross sections, virtual power plants, sharp houses, shrewd transport and keen metropolitan networks[21].

Things may include a large group of devices in the IoT context, such as heart monitoring embeds, animal biochip transponders, cameras broadcasting live streams of wild animals in sea shore front waters, sensor-operated vehicles, environmental DNA testing devices, food, warning of microorganisms, or field action devices that assist firefighters in search and rescue assignments. "Legal analysts propose "things" as a "indivisible mix of gear, programming, data and organisation. Considering the arrangement and execution of an imaginative keen energy metre over this article, proposes the arrangement and execution of an imaginative keen energy metre. The proposed adroit energy metre is based on the Internet of Things (IoT). An investigation depending on the establishment of mindful registration, learning, and broad information in the Internet of Things waa information. A data framework for making a shrewd city via the Internet of Things was proposed by Jin et al. Another worldview, called the Intellectual Internet of Things (CIoT), was created by Wu et al. to engage the existing IoT with a "cerebrum" for important level perspective. Xia et al.[14] suggested ravenous steering sans gps on 2-D and 3-D surfaces with conveyance assurance and low stretch factor. A method for misusing the data affectability of aerometric constancy for streamlining EEG detection was proposed by Rennet al. In brilliant miniature systems, Yu et al.[16] created a strategy for carbon-mindful energy cost minimization for distributed web server farms. Abdelwahab et al. spoke through distant identification of inspiring shrewd cloud administrations: a network of all empowering agents. As a non-exclusive detecting stage towards the future Internet of Things, Khan et al. explored a proposal for a reconfigurable RFID detecting tag. Zhang et al. provided medical treatment with data on Universal WSN. Främling et al. proposed a general knowledge standard from the lifecycle point of view of executives for the IoT. For green portable group detection, Sheng et al. suggested using GPS-less detection preparation. Chen et al. spoke about the combination of data to safeguard purposeful Internet of Things attacks. In the cloud-driven Internet of Things, Kantarci and Mouftah suggested dependable detection for public well-being. Lin et al. suggests a convention and a technique for the range of executives that, considering the impediments of nearby handling, should plan for normal types of

112

security hazards. New and innovative applications focused on the IoT and its nuts and bolts have been discussed in writing. A model of an energy observing gadget based on an open source concept is presented in this paper [40-41].This engineering guarantees a few focal points on traditional energy metres, such as easy enhancement of new applications, making the transition to future savvy platform foundations cost-and time-powerful, and basic acclimatisation to adjust the applicable guidelines. The advancement of the enlistment form energy metre is talked about [34]. In this article, estimation hardware is implemented for the alignment of energy metres. Its framework and metro logical representation are discussed. Two online computerised remuneration systems have been recognised and tend to boost their exhibits without expanding their costs: one constructs the phantom virtue of test signs and one modifies the transducer recurrence reaction. Exploratory findings relating to metro logical representation have shown that the calibrator so recognised is ideal for the energy metre on-site modification Figure 2. 6.

FUNCTIONAL BLOCK DIAGRAM

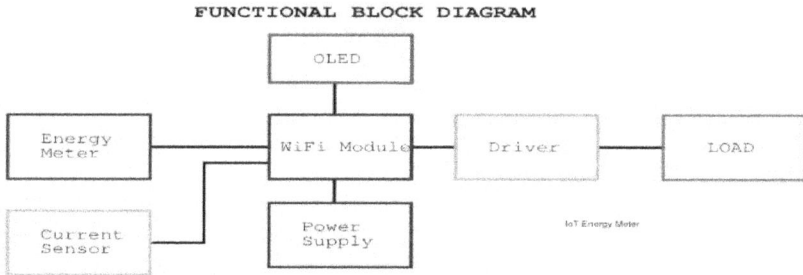

Fig 2.6 Functional block diagram of proposed IoT based energy meter

Fig 2.7 Experimental setup of IoT based smart energy meter in laboratory

In this paper, an undertaking has been made to organise and complete a practical Smart Energy Meter based on the Internet of Things model. The suggested model is used to verify the family's energy usage, and also make the analysis of the energy unit useful. Accordingly, it diminishes the wastage of energy and brings care to all. It will similarly subtract the manual interference and make Figure 2.7 savvy and trustworthy for the method.

2.5 SMART ENERGY METER READING AND MONITORING DEVICE DEVELOPMENT OF IOT BASED

The effort to collect power utility metres to peruse and discern the illegal use of power in a large portion of agricultural nations is an exceedingly troublesome and repetitive activity that requires a great deal of HR. Energy metres using the Internet of Things (IoT) to peruse and observe the system offer an efficient and financially savvy method to remotely transfer the energy data used by the customer, just as it allows offices the unlawful use of power to differentiate. The point of this research is to measure the usage of power in the family unit and ultimately establish its bill using IoT and telemetric communication procedures. In addition, this investigation means recognising and monitoring the theft of electricity. The Arduino microcontroller is used to coordinate the mechanized energy metre structure activities and to link the device to the WiFi network and thus to the Internet and the server[29].With the framework, an unsolved infrared sensor is protected to detect when any unauthorized shift happens in the metering system. In such a situation, the machine would equally give a warning to the worker as it has the workplace to detach and re-partner the force seamlessly. The suggested device is set up to screen endlessly and to notify the energy supplier and customer about the amount of units eaten. The energy uses are therefore overcome and the bill is resurrected by using an Internet of Things partnership on the internet. This computerization will minimize the need for manual work[1-5].

The requirement for appropriate energy utilization and observing mindfulness has propelled a few analysts to give inventive controlling and checking answers for the energy areas. Likewise, a few organizations give Enterprise Energy Management (EEM) programming applications to dissect the gathered information. By summing up those practices, an overall framework engineering for energy checking utilizing IoT can be resultant, as appeared in Figure 2.8. At the base layer of this design, there are keen meters and sensors, which might be associated through wired or remote organizations. Savvy energy meters accessible available can accomplish a few boundaries (for example power utilization, max/min of pinnacle voltage and force factor), thus they give a significant level of adaptability in observing and investigating energy utilization. At the mid layer, gathered information are shipped off a door, and afterward moved to a neighborhood PC or to the web through standard correspondences conventions, for example, the Zig Bee remote innovation. In the event that remote organizations are utilized, sensors can be much more deftly positioned all through the shop floor [10]. Finally, information on EEM programming for review is taken into account in other business frameworks, such as Building Management Systems (BMS), Advanced Production and Scheduling Frameworks (APS), Manufacturing Execution Systems (MES), Manufacturing Resource Planning (MRPII) or in essence, Enterprise Resource Planning (ERP). It is also possible to coordinate the data from excellent metering
frameworks through an administrative control and information procurement system (SCADA).

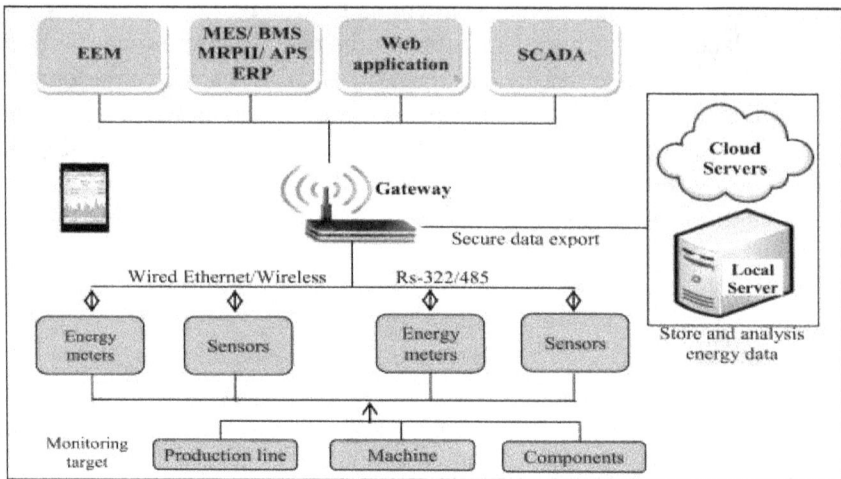

Fig 2.8 Framework for Energy Monitoring Using IoT

In the plan of keen energy meter, the microcontroller is inter-
faced with AMR module, Theft location module and Wi-Fi module. The microcontroller is a center segment of the keen energy
meter framework which is put at the customer end to quantify
the meter perusing, burglary recognition and putting away the
information. This information is moved between customer end
and energy provider end utilizing IoT ESP3866 Wi-Fi. The AMR
module ceaselessly screens the meter and gathers the perusing
and ships off the microcontroller. In the current situation, there
is a need to remarkably recognize the brilliant meter gadget distantly in a solid way. To accomplish the trait of gadget distantly
we have given IP address to every association.

2.6   Advanced Metering and Billing Warning Framework
Smart Energy Meter

Demand Side Management (DSM) will expect a fundamental portion later on a canny cross section by savvily directing burdens.
DSM programmes, recognized for sharp metropolitan zones by
Home Energy Management (HEM) systems offer buyers recognize
power confidence speculation assets and utility works at decreased
zenith interest in various favorable circumstances. This paper presents the DSM model based on Evolutionary Algorithms (EAs) (Bi-

116

nary Particle Swarm Optimization (BPSO), Genetic Algorithm (GA) and Cuckoo search) for arranging private client devices. For three examples, the model is replicated in Time of Use To U) approximate atmosphere: I normal homes, (ii) sharp homes, and (iii) Renewable Energy Sources (RES) wise homes.

Propagation findings suggest that the proposed model preferably designs the devices to decrease power bill and apexes. Force is the core reason behind every country's improvement. It has now become necessary for utilities associations to devise better, non-meddlesome, earth-safe methodology for testing the usage of utilities so that correct bills can be generated and invoiced with the snappy extension of private business, and current-day intensity customers all over the world. An amazing portion of the living and non-living objects that integrate us will be on the web in some structure in the Internet of Things (IOT) model. Driven by the omnipresence of contraptions allowed by wireless mechanical turn of events, such as Wireless Bluetooth, Radio Frequency Recognition, Wireless-Fidelity, implemented sensor, IoT has moved from its beginning stage and is really on the verge of transforming the present fixed between net into an imminent inside and out featuredInternet [31]. There are currently only around nine billion connected devices, and very close to fifty billion devices are projected to be connected by 2020[14-29]. Due to the rapid propulsion of portable correspondence innovation and the reduction of costs, there is a fusion of flexible innovation into the MSEB mechanization system. We suggest a system that collects the usage of energy from private as well as corporate zones and sends it directly to the focus worker where it is treated for bill planning on that detail. As per the correspondence medium used the AMR framework can be isolated into the wire AMR framework and remote AMR framework. In the current system for the variety of information on energy usage, the MSEB agents come and visit each person month by month, make the snap effort, and analyse the use data from the metre corporately and physically. There the preview and metre readings were checked by the official after the official and then provided to the neighborhood programming for charge estimates and bill age. We make the instalment for the received bill as a customer at that point. Too much of this cycle is a wild cycle[17]. Man-made mistakes can be incalculable. HR has been squandered and several different problems are arising. Finally, we have a long idea of creating

117

a structure that will naturally do the above cycle. The microcontroller is linked to our traditional energy metres, which will examine the metre after a fixed period of time. Remotely, these perusing metres will be submitted alongside their exceptional metre number to the integrated worker. The worker will prepare this data and deliver the bill as a result. It will ship off each client by means of SMS office after age of charge.

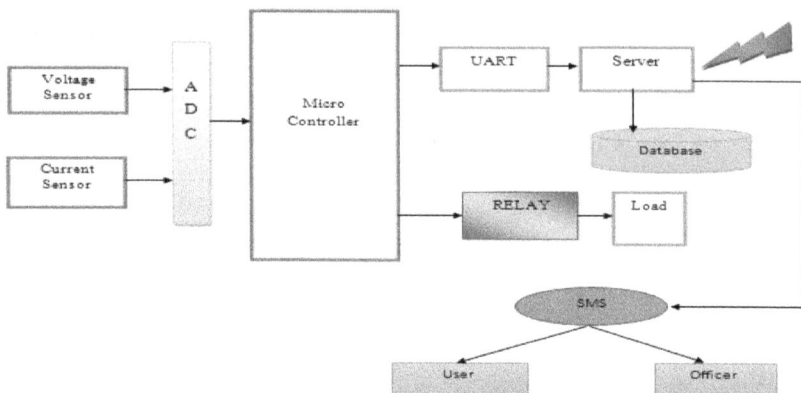

Fig 2.9 Block diagram

This paper presented another HEM model with and without RESs based on the To U estimation plan. EAs BPSO, G and Cuckoo were used in the proposed model to optimally burn-through system and RES capacity. The results of the reproductions showed that cost saving is achieved with respect to small consumer power bills. The proposed model absolutely minimized the power bill and elevated pinnacles by using BPSO, GA and Cuckoo estimates.

2.7 SMART ENERGY METERRS ANDROID BASED

The main purpose of this paper is to acquire knowledge of the layout of the standard energy metre. The calculation of used power can only be seen in the current energy metre, but the voltage, current active power, reactive power and apparent force of the structure can not be screened anyway. With the use of the splendid energy metre based on Android, all those limits can be seen without any difficulty. Its versatility, negligible effort, simple to manage and looking at the whole system are the crucial reasons for this frame-

118

work. With the critical 220V AC line that is passed on from the power line, the cycle starts. A voltage and a current sensor measure line voltage and current. These sensors are connected to the microcontroller and will check the load-related current. By then, using various mathematical states of the device, Active Power, Reactive Power, Apparent Power and Watt can be resolvedThe purposeful information will be displayed on the LCD and transmitted via Bluetooth module to the android contract and dispatched from the android device. The system's blend of hardware and programming is tested and thoroughly tried. Despite the way in which a few requirements and difficulties are encountered during the course of the endeavor, the capability of the structure is exactly as planned. In various instances, one may gain from working up an informative energy metre framework and using looming techniques. Such a gas and force metre is a savvy metre that can give metre readings to your energy supplier carefully. This will ensure more accurate bills for electricity. In addition, clever metres go with screens, so you can understand the energy consumption all the more quickly[27].

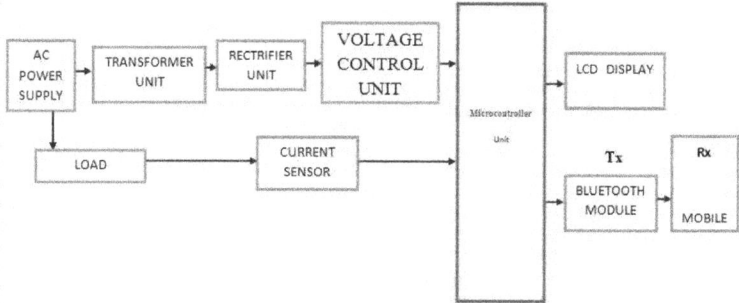

Fig 2.10. block diagram

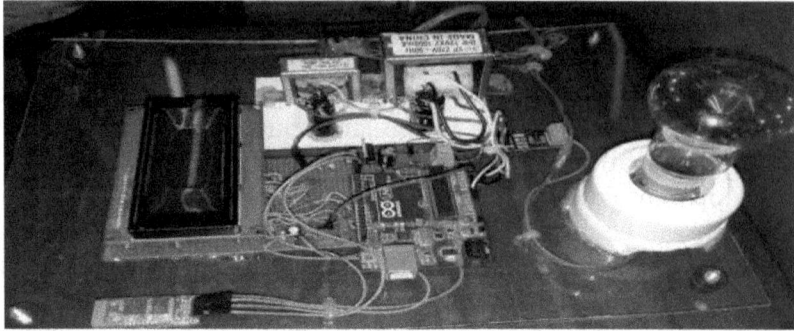

Fig 2.11 proposed system

Targets are

1.   To showcase Current, Voltage, Active force, Reactive force and clear force.

2.   Design a proficiency and easy to understand.

3.   Reduce creation cost.

4.   Power misfortune decrease reason for influence factor improvement.

5.   Wireless Meter Reading transmission.

6.   Display prompt perusing in advanced cell.

At long last, subsequent to finishing this task we get the unmistakable thought regarding the Wireless Smart Energy Meter. We examined about the segments and used to actualize this venture. We study the subtleties data about the parts used to finish this venture, for example, Microcontroller, LCD, Voltage controller, Crystal Oscillator, Zener diode and so on We come to realize how to make a circuit association and afterward check the yield. It will help us in our future work[33].

2.8 Advance energy meter using ARM micro controller

The Smart Grids are a power conveyance framework which screens,

secures and streamlines the activity of its interconnected components from start to finish. They incorporate focal and disseminated
environmentally friendly power generators through the electrical
organization. In this manner, it can undoubtedly be contended that
they are described by extraordinary issues to deal with the entire
framework and to guaranteeing a legitimate energy quality. In this
situation the likelihood to have progressed power/energy meter
with energy quality observing and correspondence capacities is of
extraordinary interest. Along these lines, in this paper the plan and
usage of a wattmeter for single stage electrical frameworks dependent on an ARM microcontroller is introduced. The wattmeter is
made out of estimation transducers (for voltage what's more, current), molding, procurement, handling and correspondence segments; the last three are installed in the microcontroller. Other than
run of the mill wattmeter errands, the framework is proficient to
quantify some force quality boundaries. In addition, correspondence
is performed through a USB based exchange square to a Host Pc,
through which it tends to be designed. Regardless of whether it is
intended to regard on-line estimation imperative, it stays an ease
framework. In the paper trial results identified with portrayal are
additionally depicted [35-40]. Electrical energy estimation assumes
a significant job in business energy exchanges as well as in the assessment of energy adjusts in ventures and in the presentation assessment of machines and energy frameworks, both customary and
creative. With the incorporated quality confirmation of the electrical
administrations the likelihood to specify contracts for quality for the
clients has been presented. This infers the obsession of a concurred
level of value ("Custom Power"), a yearly charge on the rear of the
client and a repayment for him for the situation when the degree of
value has not been regarded. This needs on-line assurance of energy
streams and a relating level of value. To arrive at this objective,
computerized signal preparing procedures can be received; these
methods are usually utilized in the present instrumentation world,
both in the logical and mechanical fields. They are primarily found-
ed on the utilization of numerical tasks. Most estimation calculations, for example, the discrete Fourier change, computerized sift-
ing, or versatile sign preparing, require the all-encompassing utilization of number juggling activities.

Also, advanced preparing of physical-world signs requires their
transformation into computerized design which can be done by

121

simple to advanced converters (ADC). Estimation of energy streams additionally requires simultaneous transformation of voltage and current signs. So at any rate two S&H and one ADC or two S&H and two ADC must be utilized, to stay away from stage mistakes between signals. Different necessities for such frameworks are compactness, correspondence capacity, to move or view the estimation results and the minimal effort, to the mean to have a spread dissemination ([1]-[3]). Instruments for on-line estimations are portrayed by a flat out time requirement to finish info, handling and yield tasks, which must not be surpassed [22-26]. They are explicitly intended to lessen: – the time spent by the information procurement framework (DAS) to get the information signal; – the time spent by the microchip to handle information; – the time spent by the correspondence interface to move estimation results. Information preparing time primarily relies upon the necessary boundaries, that is, on the embraced estimation calculation and on processor execution. A reasonable stage for these applications is the microcontroller, because of its extraordinary design that incorporates chip, perpetual memory, unpredictable memory, I/O pins, and gadgets such ADCs, correspondence interfaces, DMA, and so forth In contrast to microchips (broadly useful) a microcontroller is intended for most extreme independence and upgrading value execution proportion for a particular application. In this paper the plan and usage of a microcontroller-based wattmeter, with the on-line estimation limitation. Other than commonplace amounts estimated by a wattmeter, it can likewise assess consonant twisting. In area II there is the depiction of the acknowledged estimation instrument, its equipment engineering and estimation calculation. In area III exploratory outcomes are introduced.

The best trouble for the fashioner of instrument for on-line applications is that the DSP by and large registering force might be not sufficiently high to fulfill the time imperatives. Truth be told, aside from the DSP throughput, it must be likewise thought of if the particular prerequisites, regarding information stream, can be guaranteed by the memory and I/O structures. A few gadgets are enhanced to work with on-chip memory, regardless of whether it is in every case carefully restricted. Through a reasonable approach for memory the executives and utilizing gadgets, for example, "Direct Memory Access" (DMA), microchip has just to manage estimation calculation. This diminishes handling time, expands exhibitions, without loss of tests, keeping on-line limitations [42].

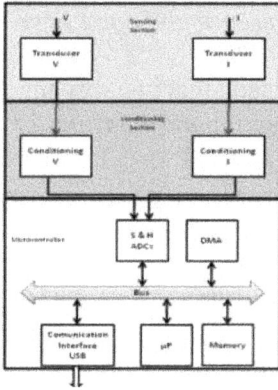

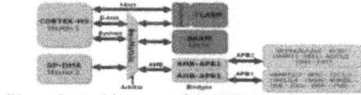

Figure 2. Architecture of the STM32 microcontroller.

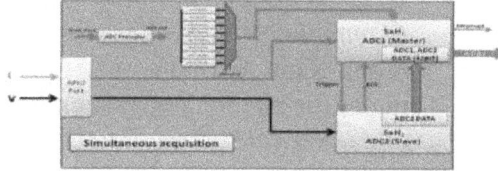

Figure 2.11 Block diagram of Arm micro-controller

CONCLUSION

This idea is being updated to decrease human dependency in order to gather perusing month by month and restrict the specialized problems with respect to charging measure. With the pre-introduction of the intensity plan using the Arduino miniature regulator and a GSM (Global System for Mobile Communication) module, this venture expands the plan and implementation of an energy observing framework. The benefit of this system is that on the daily basis, a client can realize the force consumed by the electrical devices and can find a way to manage them and thus aid in energy safety. The data regarding the bill sum, instalment and the pre-arranged force shut down subtleties are imparted to the buyer from the power board region.

References

[1]   Anitha.k,Prathik (2019), Smart Energy Meter Surveillance Using IoT,*Institute of Electrical and Electronics Engineers(IEEE),*

[2]   Devadhanishini (2019), et.al Smart Power Monitoring Using IoT5th *International Conference on Advanced Computing & Communication Systems (ICACCS).June 2019*

[3]   Mohammad Hossein Yaghmaee(2018), Design and Imple-
        mentation of an Internet of Things Based Smart Energy
        Metering *6th IEEE International Conference on Smart Ener-*
        gy Grid Engineering 2018.
[4]   Himanshu kpatel, (2018) Arduino based smart energy me-
        ter *2nd Int'l Conf. on Electrical Engineeringand Information*
        &Communication Technology (ICEEICT) 2018
[5]   Bibek Kanti Barman, et.al (2017) proposed paper smart
        meter using IoT department of *international electronics*
        and electrical engineering (IEEE) 2017
[6]   Garrab.A, Bouallegue.A, Ben Abdullah (2012), A new AMR
        approach for energy savings in Smart Grids using Smart
        meter and partial power linecommunication, *IEEE First*
        International Conference on ICICS,vol 3, pp. March 2012
[7]   Landi,c.: Dipt. Di Ing.dellInf, SecondaUniv di Napo-
        li,Aversa,Italy; Merola p (2011).ARM-based energy man-
        agement system using smart meter and Web server,*IEEE*
        instrumentation and measurement technology conference
        binjing, pp.1-5 may 2011.
[8]   B. S. Koay, S. S. Cheah, Y. H. Sng, P. H. Chong, P. Shum, Y. C.
        Tong, X. Y. Wang, Y. X. Zuo and H. W. Kuek (2003), "Design
        and implementation of Bluetooth energy meter", *IEEE Pro-*
        ceedings of the 4th International Joint Conference of the
        ICICS, vol. 3, pp. 1474-1477, Dec,2003.
[9]   N. Fathima, A. Ahammed, R. Banu, B.D. Parameshachari,
        and N.M. Naik (2017), Optimized neighbor discovery in
        Internet of Things (IoT), *In Proc. of International Confer-*
        ence on Electrical, Electronics, Communication, Computer,
        and Optimization Techniques (ICEECCOT), pp. 1-5, 2017.
[10]   M M Mohamed Musfassirin, A L Hanees (2018), Devel-
        opemnt of IoT based Smart Energy Meter Reading and
        Monitoring System, *International Conference,* 2018
[11]   Mohamed Mufassirin, M. M., &Hanees, A. L. (2018). Cost
        Effective Wireless Network
        Based Automated Energy Meter Monitoring System for Sri
        Lanka Perspective.
        International Journal of Information Technology and Com-
        puter Science(IJITCS), 68-
        75

[12]   Darshan, I. N., & Radhakrishna, K. A. (2015). IoT Based
         Electricity Energy Meter Reading,
         Theft Detection and Disconnection using PLC modem and
         Power optimization.
         *International Journal of Advanced Research in Electrical,
         Electronics and
         Instrumentation Engineering, 4(7),* 6482-6491.
[13]   Muhammed, A. A., &Hanees, A. L. (2018). IOT Based Waste
         Collection Monitoring System
         Using Smart Phones. *8th International Symposium* - 2018.
         South Eastern University
         of Sri Lanka. "In Press"
[14]   Pooja, D. T., & Kulkarni, S. B. (2016). IoT Based Energy Me-
         ter Reading. International
         *Journal of Recent Trends in Engineering & Research
         (IJRTER),* 586-591
[15]   Rishabh Jain, Sharvi Gupta, Chirag Mahajan, Ashish
         Chaulan, IoT based Smart Eneregy Meter Monitoring and
         Controlling System, Vol-7, issue-2, 2019
[16]   Pandit S (2017), "Smart energy meter using internet of
         things (IOT)", *VJER vishwakarma Journal of engineering
         Research,* Vol.1, No.2, (2017)
[17]   Giri Prasad S (2017), "IOT based energy meter", *Interna-
         tional Journal of recent trends in engineering& research
         (IJRTER),* (2017).
[18]   Sehgal VK, Panda N, Handa NR, Naval S & Goel V (2010),
         "Electronic Energy Meter with instant billing", Fourth
         UKSim *European Symposium on Computer Modeling and
         Simulation (EMS),* (2010), pp.27-31
[19]   Malik NS, kupzog F & Sonntag M (2010), "An approach to
         secure mobile agents in automatic meter reading", *IEEE
         International Conference on Cyber worlds, computer society,*
         (2010), pp.187-193.
[20]   Geetha R, Abhishek D, Rajalakshmi G, Sabari murugan,
         Surendar V (2018), *Smart Energy Meter using IoT, IJRTER,*
         ISSN:2455-1457, 2018
[21]   Himshekhar Das, L.C.Saikia (2015), "GSM Enabled Smart
         Energy Meter and Automation of Home Appliances", PP-
         978-1-4678-6503-1, 2015 *IEEE*

[22]   Ofoegbu Osita Edward (2014), "An Energy Meter Reader with Load Control Capacity and Secure Switching Using a Password Based Relay Circuit", PP-978-1-4799-8311-7, 'Annual Global Online Conference on Information and Computer Technology', *IEEE* 2014.

[23]   Yingying Cheng, Huaxiao Yang, Ji Xiao, Xingzhe Hou (2014), "Running State Evaluation Of Electric Energy Meter", PP-978-1-4799-4565-8, 'Workshop on Electronics, Computer and Applications', *IEEE* 2014.

[24]   Luigi Martirano,MatteoManganelli,DaniloSbordone (2015),"Design and classification of smart metering systems for the energy diagnosis of buildings" *IEEE* 2015.

[25]   J. Widmer, Landis, (2014)" Billing metering using sampled values according to lEe 61850-9-2 for substations",*IEEE* 2014.

[26]   Cheng Pang,ValierryVyatkin,Yinbai Deng, Majidi Sorouri (2013), "Virtual smart metering in automation and simulation of energy efficient lighting system" *IEEE* 2013.

[27]   Abhiraj Prashant Hiwale, Deepak Sudam Gaikwad, Akshay Ashok Dongare, Prathmesh Chandrakant Mhatre (2018), Iot Based Smart Energy Monitoring, *IRJET*, vol-05, issue-03, 2018

[28]   B. S. Koay, S. S. Cheah, Y. H. Sng, P. H. Chong, P. Shum, Y. C. Tong, X. Y. Wang, Y. X. Zuo and H. W. Kuek (2012)", "Design and implementation of Bluetooth energy meter", 2012

[29]   Darshan Iyer N, Dr. KA Radhakrishnan Rao, (2015) "IoT Based Energy Meter Reading, Theft Detection & disconnection using PLC modem and Power optimization", *IRJET*, (2015)

[30]   Dr. Aditya Tiwary, Manish Mahato, Mohit Tripathi, Mayank Shrivastava, Mayank Kumar Chandrol, AbiteshChidar (2018), Design and Implementation of an Innovative Internet of Things (IOT) Based Smart Energy Meter, *IJFRCSCE*, vol-4, issue-4, 2018

[31]   V. Ovidiu, F. Peter (2013), "Internet of Things: Converging Technologies for Smart Environments and Integrated Ecosystems", Aalborg, Denmark: *River Publishers*, 2013.

[32]   M. Friedemann, F. Christian "From the Internet of Computers to the Internet of Things", ETH Zurich.

[33]   W. Na, J. Park, C. Lee, K. Park, J. Kim, and S. Cho (2018), "Energy-Efficient Mobile Charging for Wireless Power Transfer in Internet of Things Networks", *IEEE Internet of Things journal*, Feb. 2018, Vol. 5.

[34]   Q. Wu, G. Ding, Y. Xu, S. Feng, Z. Du, J. Wang, K. Long (2014), "Cognitive Internet of Things: A New Paradigm Beyond Connection", *IEEE Internet of Things journal*, April 2014, Vol. 1, pp. 129-143

[35]   S. Abdelwahab, B. Hamdaoui, M. Guizani, A. Rayes (2014), "Enabling Smart Cloud Services Through Remote Sensing: An Internet of Everything Enabler", *IEEE Internet of Things journal*, June 2014, Vol. 1, pp. 276-288.

[36]   K. Framling, S. Kubler, A. Buda (2014), "Universal Messaging Standards for the IoT From a Lifecycle Management Perspective", *IEEE Internet of Things journal*, June 2014, Vol. 1, pp. 319-327

[37]   S. A. Joshi, S. Kolvekar, Y. R. Raj, S. Singh (2016), "IoT Based Smart Energy Meter", *International Journal of Research in Communication Engineering*, Vol. 6, 2016

[38]   Gobhinath S, Gunasundari N, Gowthami P (2016), "Internet of Things (IOT) Based Energy Meter", *International Research Journal of Engineering and Technology (IRJET)*, Vol. 3(4), 2016.

[39]   Ashna.K and Sudhish N George (2013), "GSM based automatic energy meter reading system" *IEEE Wireless communications*, 2013.

[40]   O. Vermesan, P. Friess, P. Guillemin (2011), "Internet of things strategic research roadmap," *Internet of Things: Global Technological and Societal Trends,* vol. 1, pp. 9–52, 2011

[41]   O. Said, M. Masud,(2013) "Towards internet of things: survey and future vision," *International Journal of Computer Networks*, vol. 5, no. 1, pp. 1-17, 2013

[42]   A. Tiwary, M. Mahato, M. Tripathi, M. Shrivastava, M. K. Chandrol, A. Chidar,(2018) "A Comprehensive Review of Smart Energy Meters: An Innovative Approach", *International Journal on Future Revolution in Computer Science & Communication Engineering*, Vol. 4, No. 4, April 2018

Chapter 8

Investigation of Metallurgical Characteristics with Super Duplex Stainless Steel Plates

Madhu M.C. & Prashanth N.A.
Department of Mechanical Engineering, BMS Institute of Technology and Management, Bengaluru, India

Abstract

Steel is a composite material made up of iron, carbon, and other components. It is a key component used in structures, infrastructure, tools, ships, vehicles, machineries, appliances, and weaponry because to its great tensile strength and inexpensive cost. The use of Super Duplex Stainless steel in transportation, such as automobiles, such as the Exhaust silencer for Mercedes Benz CL III coupe and Propeller shaft, and also in marines, necessitates the use of more efficient and reliable welding techniques and precise welding parameters in order to achieve strong joints and extend the life of materials. The goal of this study is to assess the impact of activated tungsten inert gas welding settings on the metallurgical characteristics of Super duplex stainless steel plates with dimensions of 200 mm length, 150 mm width, and 5 mm thickness. The welding parameters under consideration were the Welding Current, Voltage, and Gas flow rate. Optical microscope and scanning electron microscope tests were used to evaluate the metallurgical characteristics with welding settings. We analyzed the microstructure of the welded specimen. Welding strength and durability were evaluated as a result of this technique.

1. INTRODUCTION

Carbon may make up to 2.14 percent of the weight of standard steel alloys. Varying the amount of carbon and many other alloying elements in the final steel, as well as controlling their chemical and physical make-up (either as solute elements or as precipitated phases), slows the movement of those dislocations that make pure iron ductile, and thus controls and enhances its qualities. These characteristics include hardness, quenching behavior, annealing requirements, tempering behavior, yield strength, and tensile strength of the resultant steel [1]-[5]. Because the crystal structure of pure iron provides minimal barrier to atoms sliding past one another, pure iron is ductile, or soft, and easy to produce. Small quantities of carbon, other elements, and inclusions in the iron function as hardening agents in steel, preventing the movement of dislocations that are typical in iron atom crystal lattices.

Modern steels are created from a variety of alloy metals to serve a variety of purposes. Carbon steel, which is made up entirely of iron and carbon, accounts for 90% of all steel manufacturing. To increase the harden ability of thick sections, low alloy steel is alloyed with additional elements, commonly Molybdenum, Manganese, Chromium, or Nickel, in quantities of up to 10% by weight [6]-[10]. To give increased strength for a little price increase, high strength low alloy steel incorporates tiny additions (generally less than 2% by weight) of other elements, often 1.5 percent manganese. A new kind of steel called as Advanced High Strength Steel has emerged as a result of recent Corporate Average Fuel Economy (CAFE) standards (AHSS). This material is robust and ductile, allowing vehicle structures to maintain their present levels of safety while utilizing less material. There are various commercially available grades of AHSS, such as dual-phase steel, which is heat-treated to generate a formable, high-strength steel with both a ferric and martens tic microstructure. TRIP steel uses unique alloying and heat treatments to stabilize austenite concentrations at room temperature in generally austenite-free low-alloy ferric steels. The austenite undergoes a phase change to marten site without the input of heat when strain is applied. Twinning Induced Plasticity (TWIP) steel uses as Specific type of strain to increase the effectiveness of work hardening on the alloy. For rust resistance, carbon steels are often galvanised by hot-dip or electroplating in zinc [11]-[15].

2. RELATED WORK

In the year 2014, investigate the effects of welding settings on TIG welding of aluminum plate. TIG welding was used to improve the welding properties of aluminium plates, and a study was conducted to determine the parameters obtained. In the year 2015, A Review on TIG Welding for optimizing process parameters on dissimilar joints Tungsten Inert Gas (TIG)Welding is used to join heat treatable stainless steel (SS 304) and mild steel (MS 1018). In the year 2016, researchers looked at the effects of TIG Welding process parameters on Mechanical properties. TIG welding of stainless steel (SS 304) material and investigation of parameters after welding The welding process parameters are carefully monitored to determine the outcome. In 2017, a review of the parametric optimization of TIG welding was published. On TIG welding, optimizing welding parameters such as welding speed, material flow rate, and so on. In the year 2017, a novel arc heat treatment approach was used to develop the microstructure and functionality of a specifically graded Super Duplex Stainless Steel. It has been claimed that the optimal attribute of SuperDuplexStainlesssteel is a balanced ratio of Ferrite and Austenite Structure. In the year 2014, researchers evaluated the microstructure and mechanical characteristics of a Dissimilar Austenitic/Super Duplex stainless steel joint. Large Austenite grains with small levels of Ferrites in Super Duplex base metal make up the heat-affected zone of Austenitic base metal. Structures, characteristics, and uses of super and hyper duplex stainless steels in 2016. Among current duplex stainless steels, Hyper Duplex Stainless Steels have the greatest Critical pitting temperature and Critical crevice corrosion temperature. In the year 2018, ferrite content meter analysis for delta ferrite evaluation in super duplex stainless steel It is feasible to apply the calibration pattern offered by ferritoscope manufacturers for stainless steel with low ferrite concentration (up to 35 percent ferrite). During the drilling process, a mach inability study of first-generation duplex (2205), second-generation duplex (2507), and austenite stainless steel was reported in 2013.

The Effect of Constant and Pulsed Current Gas Tungsten Arc Welding on Joint Properties of 2205 Duplex Stainless Steel to 316L Austenitic SS published on 2016 discussed dissimilar 316 Laustenitic-steel/2205 duplex stainless steel (DSS) joints were fabricated by constant and pulsed current gas tungsten arc welding using ER2209 DSS as a filler metal. In 2016, a paper titled Mechanical characterization of gas tungsten arc welded super duplex stainless steel joints was released. Super duplex stainless steels (SDSS) are employed in high-tech applications in the construction, chemical, and shipbuilding sectors due to their exceptional characteristics. During welding, however, SDSS suffers micro structural changes that cause the ferrite-to-austenite ratio to be unbalanced. Eddy current approach for characterization of super duplex stainless steel. Several SDSS samples with low s phase concentration and non-balanced microstructure were purposely generated by thermally processing SDSS specimens, according to a study published in 2015. To analyze the SDSS samples, electromagnetic methods such as standard Eddy Current Testing (ECT) and Saturated Low Frequency Eddy Current (SLOFEC) were used.

3. PROPOSED WORK

Super Duplex steels, such as SS32750, have a blended microstructure of austenite and ferrite (50/50), giving them more strength than ferric and austenitic steels. The key distinction is that Super Duplex has a higher molybdenum and chromium content than ordinary duplex grades, giving it more corrosion resistance. In high-chloride settings, the balanced dual phase microstructure combines excellent strength with cost-effective corrosion resistance. Due to the material's improved tensile and yield strength, Super Duplex offers the same advantages as its predecessor. It has cheaper alloying costs when compared to equivalent ferric and austenitic grades with equipment corrosion protection in chloride containing environments. In many circumstances, this allows the buyer to acquire lesser thicknesses without sacrificing quality or performance.

Table. 1. Mechanical Properties of SDSS

Tempered	Annealed
Tensile Rm	115 ksi (min)
Tensile Rm	800 Mpa (min)
R.p. 0.2%yield	80 ksi (min)
R.p. 0.2%yield	550 Mpa (min)
Elongation (2" or 4 Dgl)	15 % (min)
Hardness	Rc32 max

Table. 2. Physical Properties (Room Temperature)

Tempered	Annealed
Specific Heat (0-100°C)	500 J.kg-1°K-1
Thermal Conductivity	15 W.m-1.°K-1
Thermal Expansion	11 μm/μm/°C
Modulus Elasticity	200 GPa
Electrical Resistivity	8.12 μohm/cm
Density	7.8 g/cm3
Melting Point	1410/1460 °C

Table.3. Chemical Composition (%by weight)

Element	Min	Max
C	-	0.03
Cr	24	26
Cu	0.5	1
Mn	-	2
Mo	3	5
N	0.24	0.35

Ni	6	8
P	-	0.35
S	-	0.015
Si	-	1

Welding is a manufacturing or artistic method that joins materials, often metals or thermoplastics, by producing fusion, as opposed to lower-temperature metal-joining processes like brazing and soldering, which do not melt the base metal. A filler material is routinely added to the junction in addition to the base metal to generate a pool of molten material (the weld pool) that cools to form a joint that is usually stronger than the base material. Pressure may be used alone or in combination with heat to create a weld. Welding also necessitates the use of a shield to prevent contamination or oxidation of the filler metals or molten metal's. Welding may be done using a variety of energy sources, including a gas flame, an electric arc, a laser, an electron beam, friction, and ultrasound. Welding may take place in a variety of locations, including the open air, beneath water, and in space. Welding is a dangerous job, and safety measures must be taken to prevent burns, electric shock, visual loss, inhalation of noxious gases and fumes, and exposure to extreme UV radiation. To melt metals at the welding site, these procedures require a welding power source to produce and sustain an electric arc between an electrode and the base material. They can work with direct current (DC) or alternating current (AC), as well as consumable and non-consumable electrodes. A shielding gas, or inert or semi-inert gas, is occasionally used to protect the welding area, and filler material is sometimes employed as well.

Oxyfuel welding, commonly known as oxyacetylene welding, is the most popular gas welding procedure. Although it is one of the oldest and most versatile welding methods, it has lost favour in industrial applications in recent years. It is still commonly used for repairing and welding pipes and tubes. The equipment is very affordable and straightforward, often relying on the combustion of acetylene in oxygen to generate a welding flame temperature of about 3100 degrees Celsius. Because the flame is less focused than an electric arc, it causes slower weld cooling, which may result in higher residual stresses and weld distortion. However, it makes welding high alloy steels easier. Metals are cut using a similar method known as oxy-fuel cutting.

The formation of heat by flowing current through the resistance generated by the contact between two or more metal surfaces is known as resistance welding. As a large current (1000–100,000 A) is sent through the metal, little pools of molten metal develop at the weld location. Resistance welding procedures are generally efficient and pollutant-free, but their applications are restricted and equipment costs might be costly. Energy beam welding methods, including as laser beam welding and electron beam welding, are relatively recent techniques that have grown in popularity in high-volume manufacturing. The two processes are quite similar, with the exception of the power source. Laser beam welding employs a highly focused laser beam, while electron beam welding is done in a vacuum and uses an electron beam.

4. RESULTS AND DISCUSSION

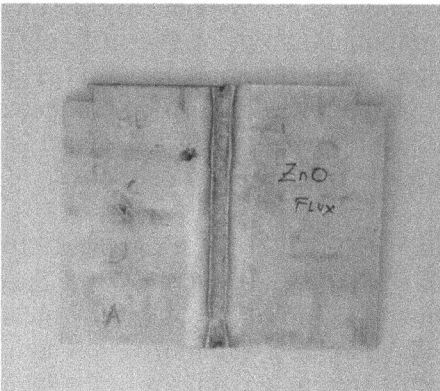

Figure. 1. Appearance o f A-TIG Welding
Welding Current: 80-150 Amp, Arc Voltage: 10-12 V, Welding speed: 40-80mm/Min, Gasflowrate: 13-18l/Min, Groove Angle: 60 Degree, Welding Current: 80-150 Amp, Welding Current: 80-150 Amp, Arc Voltage: 10-12 V, Welding speed: 40-80mm/Min

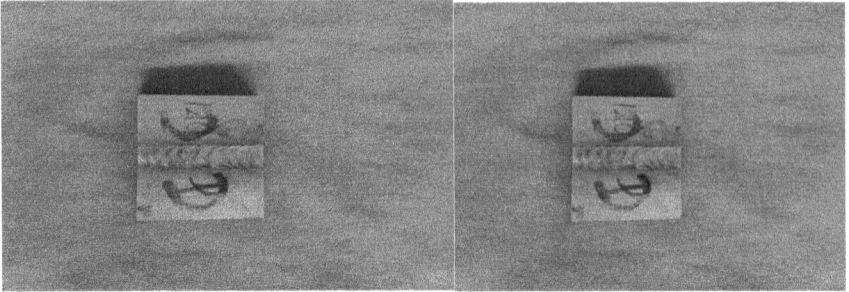

Figure.2. Material is machined through EDM for microstructure analysis

Material is machined to precise dimensions, such as 200 mm in length, 150 mm in width, and 5 mm in thickness, and then welded using A-TIG welding, as illustrated in Figure.1. Material is machined for metallurgical characteristics analysis using E.D.M, as indicated in Figure. 2.

When using the EDS feature, the signals created during SEM analysis provide a two-dimensional picture that reveals information about the sample, including exterior morphology (texture), chemical composition, and orientation of the components that make up the sample. To identify components in or on the surface of the sample for qualitative information, the EDS component of the system is used in combination with SEM analysis. It also determines elemental composition for semi-quantitative findings and detects non-organic foreign compounds and coatings on metal.

LTI uses a Hitachi S-3400NSEM for scanning electron microscopy. With an EDAX EDS, the Energy Dispersive X-ray Spectroscopy component is implemented. The SEM equipment includes a variable pressure system that can hold wet and/or non-conductive samples with minimum pretreatment. Samples up to 200 mm (7.87 in.) in diameter and 80 mm (3.14 in.) in height may be examined in the large sample chamber. SEM analysis produces high-resolution pictures at magnifications ranging from 5x to 300,000x. The EDS features a silicon drift detector (SDD) that offers. Superior speed and energy resolution compared with traditional Si Li detectors. The system is driven by the new TEAM software package, which enables material characterization using techniques including spectrum analysis, line scanning, and element mapping.

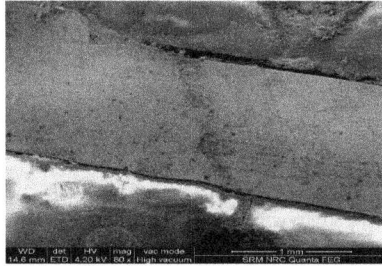

Figure.3. shows the SEM image of the heat input of 100A welded specimen of the magnification under80

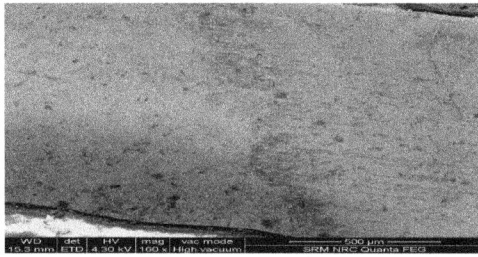

Figure.4. shows the SEM image of the heat input of 100A welded specimen of the magnification under160

* SEM Analysis with EDS–qualitative and semi-quantitative results.
* Magnification–from 5xto300, 000x.
* Sample Dimensions: up to 200 mm (7.87 in.) in diameter and 80 mm (3.14 in.) in height
* Solid inorganic materials, such as metals and polymers, are analyzed.
* Working distance(WD) - is calculated electronically as the distance between the point of focus on the sample surface and the SEM column above.
* A secondary electron and back-scattered detector is an Everhart-Thornley detector (ETD).
* One of the versalite high resolution scanning electron microscopes is the high vacuum (HV).
* Vaccum mode-vacuum mode whether High vacuum or Low vacuum.

5. CONCLUSION

The effect of the Activated TIG welding technique on Super Duplex Stainless Steel UNSS 32750 has been investigated using various process parameters. The following conclusions may be drawn. Material is welded using various heat inputs such as 60 A, 80 A, 100 A, and 120 A, and its microstructure is examined using an optical microscope. Conducted a SEM test on 100 A heat input welded specimen and analyzed its structure. At ideal circumstances, grain size will be finer, and larger heat inputs will lead to grain expansion, which will result in a loss of hardness characteristics. Reducing rain boundaries increases the possibility and quantity of dislocation movement as structural line faults. Welded metal will lose strength and hardness as a result.

REFERENCES

[1] Li HL, Liu D, Yan YT, Guo N, Feng JC.Microstructural characteristics and mechanical properties of underwater wet flux-cored wire welded 316L stainless steel joints(2016).*Journal of Materials Processing Technology*.2016,238:423-430.
[2] Wu Hongbo, Wang Zhidong, Xiang Bing (2012). Domestic and international nuclear power development situation analysis [J]. *Energy Technology Economy* .2012 (03): 5-9.
[3] Wang Hai-tao, Shan Jian-qiang, Gou Jun-li, Zhang Bo (2015). Study on accident of LOCA condition in spent fuel pool [J]. *Nuclear Power Engineering* .2015 (04): 136-139.
[4] Feng Yunliang. Nuclear spent fuel pool underwater local dry robot welding power supply [D]: *South China University of Technology*.
[5] Wang Zhenmin, Feng Yunliang (2017). Underwater robot local dry welding system [J]. *Welding Journal* .2017,38 (1): 5-8.
[6] Wu Kaiyuan, Huang Shisheng, Li Yang, Lu Peitao (2005). Digital control technology of arc welding power supply [J]. *Semiconductor Technology* .2005 (01): 72-75.
[7] Gao Hongyu. Underwater welding wire feeder development and underwater welding technology experimental study [D]: *Harbin Institute of Technology*.

[8] Zhu Jialei. Nuclear power plant maintenance of local dry automatic underwater welding technology research [D]: *Beijing University of Chemical Technology*.

[9] Duan Yu, Chen Xiao-qiang, Tang Yao. Application of New Underwater Local Dry Welding (2013) *Metalworking (Hot Working)* .2013 (12): 46-48.

[10] Ikeda, Rinsei; Okita, Yasuaki; Ono, Moriaki; Yasuda, Koichi; Tera- saki, Toshio (2010). *Quarterly Journal of the JWS*. 2010, vol. 28, no. 1, p. 141–148.

[11] Taniguchi, Koichi; Ikeda, Rinsei; Oi, Kenji. *Guide of the National Meeting of JWS*. 2012, vol. 90, p. 240–241.

[12] Sadasue, Teruki; Igi, Satoshi; Taniguchi, Koichi; Ikeda, Rinsei; Oi, Kenji (2014). *Quarterly Journal of the JWS*. 2014, vol. 32, no. 2, p. 64–72.

[13] Matsushita, Muneo; Ikeda, Rinsei; Oi, Kenji (2013). Guide of the National *Meeting of JWS*. 2013, vol. 92, p. 223–224.

[14] Yoshikawa, Nobuhiro; Tarui, Taishi; Mori, Kiyokazu; Sakamoto, Takeshi (2010). Proceedings of the 73rd Laser *Materials Processing Conference*. 2010, no. 73, p. 53–56. [15] Kobayashi, Sigeru; Gomi, Tetsuya. *Honda R & D Technical Review*. 2010, vol. 22, no. 1, p. 188–193.

[16] Suzuki, Shinichi; Oi, Kenji; Itimiya, Katsuyuki; Kitani, Yasushi; Murakami, Yoshiaki (2004). *Materia Japan*. 2004, vol. 43, no. 3, p. 232– 234

Chapter 9

Recent Trends of Welding Technology and Applications

Madhu M.C
Department of Mechanical Engineering, BMS Institute of Technology and Management, Bengaluru, India

Abstract

Welding is one of the most essential and adaptable manufacturing methods accessible to industry. Welding is used to connect a variety of commercial alloys in a variety of forms. Many items, such as guided missiles, nuclear power plants, jet aircraft, pressure vessels, chemical processing equipment, transportation vehicles, and literally hundreds of others, could not be created without welding. Many of the issues that come with welding may be avoided by paying attention to the process's unique features and needs. The joint must be designed correctly. Understanding the enormous number of accessible choices, the diversity of potential joint configurations, and the various factors that must be defined for each operation is required for selecting the particular method. Tungsten inert gas welding is a common method for connecting ferrous and nonferrous metals. When opposed to MIG welding, TIG welding has a number of benefits, including the ability to connect dissimilar metals, a low heat affected zone, and the lack of slag. The kind of power supply (DCSP, DCRP, or ACHF), welding speed, and the type of inert gas used for shielding all have a role in the precision and quality of welded connections. Steel materials must have very high tensile strength and be able to cope with complicated structures of high-performance components in such situations. The trend of mega-structural building and high-efficiency transportation leads to a rigorous demand for thick and high tensile strength steel products in the area of thick plates and steel pipes. To effectively exploit such improved steel products, welding technology advancements are required, and numerous welding technologies have been created and deployed in conjunction with the advancement of steel materials.

1. Introduction

Steel materials are crucial, but advancements in welding processes that make optimal use of sophisticated steel materials are also essential. High tensile strength steels have gradually replaced mild steel in recent years, as seen by the rising use of ultra-high strength steels, and the trend of this innovation has been to manage compensated formability and weldability of chosen materials1). To make the greatest use of these high tensile strength steels and build a worldwide position of technical dominance, progress in welding methods employing these materials is urgently needed. In the automotive industry, weldability of high tensile strength materials by various welding methods, such as resistance spot welding, arc welding, laser welding, and so on, must be on par with mild steel, while joint strength corresponding to the higher strength of the base metal must be ensured, and corrosion resistance, crack resistance, and other properties must be satisfied in the same way.

If the potential advantages of welding are to be realised while detrimental side effects are avoided, careful thought should be paid to the process selection and joint design. In general, process factors during the welding process have a significant impact on the quality of a weld connection. A suitable selection of process factors should be used to generate high-quality welds, which in turn leads to bead geometry optimization. The purpose of this research is to look into the influence of welding speed and joint design factors on the tensile strength of welded joints. Several test specimens were welded with varied welding speeds (using an adjustable speed motor) and a range of alternative joint shapes to achieve this goal (bevel angel and bevel height). These investigations demonstrated that a number of GTAW process factors, as well as a variety of conceivable joint configurations, had a significant impact on the weld joint's tensile strength.

In the sector of thick plates, high strength/heavy gauge steel goods are becoming more popular as a result of the growing popularity of bigger container ships and higher high-rise structures. Following these trends, high efficiency welding methods with high heat input for high strength, thick gauge plates such as YP460 in shipbuilding and HBLTM 385 and HBLTM 440 in construction have been developed in recent years2, 3). Although thick materials of 80–100 mm in thickness have lately become popular, the traditional 1-pass high heat input welding technology is still unable to meet joint characteristics and weldability requirements. As a result, it is necessary to create novel, low-heat-input welding processes as well as high-performance welding consumables.

2. Welding Technologies

The manufacturer's issue is controlling the process input parameters in order to get a satisfactory welded junction with the requisite weld quality. To get a welded connection of the requisite quality, it has always been important to examine the weld input parameters for welded products. To accomplish so, a time-consuming trial-and-error development technique is required. The welds are next checked to see whether they match the requirements. Finally, weld settings may be selected to generate a welded junction that nearly matches the joint properties. Also, since welds may be generated with many diverse parameters, an optimal welding parameter combination is not always attained or evaluated. To put it another way, there is often a better welding input parameter combination that may be employed. Various optimization strategies may be used to describe the desired output variables by constructing mathematical models to characterise the connection between the input parameters and output variables in order to solve this challenge. To carry out this optimization, design of experiment (DoE) approaches were used. The Taguchi technique has been developed for a variety of applications in many fields.

There are various types of friction welding procedures, including rotary friction welding (RFW), linear friction welding (LFW), and the most recent, friction stir welding (FSW). The essential idea, however, is always the same: to get the junction, heat the materials to a plastic state and then employ "upsetting force" to plastically displace the materials and make the weld. Friction between two welded components (RFW and LFW) or between components and a properly designed tool generates heat (FSW). Welding operations are categorised as solid-state joining methods since they do not involve the melting of connected materials.

Many technical and economic benefits of these joining techniques, such as high process efficiency and stability, or better occupational

143

safety and health conditions than conventional welding technologies, contribute to their popularity. However, the ability to combine materials with various qualities seems to be the most significant lately. Because intermetallic compounds are produced in the fusion zone between two distinct materials, the connecting process of dissimilar materials is sometimes quite challenging. It is vital to understand and study the phase diagram of the two welded materials in order to create a high-quality junction. In addition, the microstructure and other characteristics of intermetallic phases, such as fracture sensitivity, ductility, and corrosion resistance, are critical. There are several other aspects to consider, such as the coefficients of thermal expansion of welded materials and their melting temperatures, which must be understood when welding different materials.

Friction welding, often known as forge welding, is a kind of solid state welding that involves the application of friction and pressure between two matching metal surfaces. By rubbing two metals against one other, the required heat may be created, and the temperature can be raised to the point where the portions exposed to the friction can be welded together. Friction welding is a group of solid-state welding methods in which heat is generated by mechanical friction between moving and stationary work parts, with the addition of an upsetting force to plastically displace the material.

The conversion of mechanical energy to thermal energy is the basic mechanism behind this process. One work piece rotates around its axis, while the other to be welded to it is fixed and does not spin, but may be adjusted axially to make contact with the revolving work piece. At the moment of fusion, the rotation is halted, and forging pressure is applied axially to the stationary work piece.

3. Advanced Welding Technologies

Grain structure is refined as a result of heat work. Then, without melting the parent metal, welding takes place. The majority of materials in industry are manufactured into the required shape using one of the following methods: casting, forming, machining, or welding. The choice of a process is influenced by a number of criteria, including the component's form and size, the level of accuracy needed, the cost of the material, and its availability. To attain the required outcome, one particular technique may be applied. However, it is more frequently than not feasible to choose from a variety of end-product manufacturing procedures. Among the numerous possibilities, the economics is the most important factor to consider before making a final decision. Some metal components need to be protected against corrosion. All corrosion resistant alloys are more costly than low carbon steels. In certain applications, corrosion resistance is only required on the material's surface. As a result, cladding low carbon steel with ASS wire is a cost-effective option. Cladding may save you up to 80% on the price of solid alloy. Due to its greater deposition rate and deeper penetration, gas metal arc welding (GMAW) is widely employed for manufacturing. Cladding and surfacing are also done using gas metal arc welding. The influence of GMAW parameters on cladding parameters has been explored and investigated by a number of researchers.

Duplex stainless steels are based on the iron-chromium-nickel system and are two-phase alloys. In their microstructure, these alloys typically include equal quantities of the body centred cubic (BCC) ferrite and face centred cubic (FCC) austenite phases, as well as additions of molybdenum, nitrogen, tungsten, and copper. Nickel level often varies from 5% to 10% by weight, whereas chromium concentration typically ranges from 20% to 30% by weight. Duplex stainless steels consolidate as 100 percent ferrite, but when cooled to temperatures over 1040°C, around half of the ferrite converts to austenite. In contrast to austenitic grades, this characteristic is achieved by increasing Cr and lowering Ni concentration.

Because of their superior mechanical qualities, duplex stainless steels are increasingly being employed as structural materials in construction and architecture. Their solution annealed yield strength is more than double that of normal austenitic stainless steels that aren't alloyed with nitrogen at room temperature. They have become more significant in the building of bridges in recent years, particularly when special environmental circumstances are combined with the necessity for strong load-bearing capacity. The combination of high strength and corrosion resistance makes duplex stainless steels a popular choice. Where the material comes into touch with salt water, large concentrations of chlorides are present in the ambient air, or de-icing salts are a problem, their full potential is utilised. Despite their advantageous qualities, they still fail at the weld spots, which might be due to the welding technique, process variables employed, or welding environmental pollution. As a result, the goal of this study is to see how the welding technique and heat treatment affect the mechanical characteristics of duplex stainless steels. Tensile strength, impact strength, and hardness are the mechanical qualities to be researched, since we already know that the endurance of a welded structure is directly dependent on the ensuing mechanical properties after welding. The impact of stress relief, quenching, and hardening heat treatment procedures on the mechanical characteristics of duplex stainless steel will also be investigated.

Welding is a manufacturing method that involves fusing the surfaces of the pieces to be joined together with or without the use of pressure and a filler substance to form a permanent bond. The materials that must be bonded may be comparable or different. The heat necessary for material fusion may be supplied by gas combustion or an electric arc. Because of the faster welding speed, the latter approach is more often utilised. Welding is widely utilised in fabrication as a substitute for bolted and riveted connections and as an alternative to casting or forging. It's also employed as a repair medium, for example, to reassemble a metal at a fracture, to rebuild a tiny component that's broken off, such as a gear tooth, or to repair a worn surface, such as a bearing surface.

MIG's Mechanical Characteristics Heat input has an influence on welded dissimilar joints. The heat input is determined by the welding current, voltage, and wire speed. The foundation material was IS2062, IS45C8, and IS103Cr1. A filler wire of 1.2 mm diameter copper plated mild steel was employed. When the heat input was raised, both joints (IS2062 & IS45C8) and (IS2062 & IS103Cr1) increased the tensile strength, and when the heat input was lowered, both joints (IS2062 & IS45C8) and (IS2062 & IS103Cr1) increased the hardness value. Increase the quality and productivity of weldments by optimising gas metal arc welding process parameters. Weld dilution was used as an output parameter in this study to improve weld quality and productivity, and the influence of input parameters such as wire feed rate (W), welding voltage (V), nozzle-to-plate distance (N), welding speed (S), and gas flow rate (G) was discovered. The basis material for the experiment is ST-37 steel plate, while the shielding gas is a combination of 80% argon and 20% CO_2. Taguchi's L25 orthogonal array was used to construct the experi-

146

ment, and ANOVA was used to analyse the data. They also developed a mathematical model for weld dilution. They discovered that the wire feed rate has the most significant influence on weld dilution, whereas the gas flow rate has no effect on weld dilution, based on the experimental results. Taguchi's technique and Grey Relational Analysis were used to explore the effects of welding current, wire diameter, and wire feed rate on weld bead hardness for MIG and TIG welding (GRA). The welding current was shown to be the most important characteristic for MIG and TIG welding in the research. The ideal parameter combination for MIG welding was identified using the GRA optimization approach to be welding current, 100 Amp; wire diameter, 1.2 mm; and wire feed rate, 3 m/min. The advancement of manufacturing processes in maritime engineering and historical metalworking procedures are connected to the development of the welding process. Due to advancements in metallurgy, the creation of electricity, and the development of industrial gases, weldability became a reality. Metal fusion welding is the method of connecting metals using a gas or electric arc. These techniques were invented around the end of the nineteenth century and followed quite distinct adoption and dissemination paths. The discovery of ways for producing oxygen and later acetylene, and the combination of these two, permitted the development of cutting and welding technologies, led to the creation of gas welding.

The poor penetration of electricity in many workplaces and the sluggish adoption of new production processes were two problems that impeded the usage and growth of electric arc welding. The employment of these new production processes to repair warships and the building of an all-welded steel structure for a factory were forerunners to the transition away from using welding as a repair technology. With the conclusion of World War II, drastic changes were ushered in, resulting in a massive development of new industries that replaced archaic techniques. The utilisation of technology innovation is well-known for producing better goods at cheaper prices, and technological innovation provides higher downstream advantages for both current and new consumers. Welding is a manufacturing procedure that is utilised in a wide range of industries, and it is estimated that welding technology supports over two million employment in Europe. Welding has an economic effect since it is employed in the manufacturing industry, which creates around 1600 billion euros in added value each year in Europe. Joining technology is anticipated to account for around 5% of total value, or 8 billion euros. The process used to weld pipes or tubes is known as orbital welding. When it comes to pipelines, this technology is employed on a huge scale since a pipeline is made up of a number of pipes that are connected together. There is a danger that these pipes may leak at their junctions. These leaks might be dangerous if the pipes are used to transport gas or any other hazardous liquid, such as acid. Leakage may also result in a decrease in production. Welding technology must develop in order to solve these issues. As the utilisation of pipelines grows, the need for orbital welding grows as well. Tube welding, like pipe welding, has become a fundamental industry necessity in recent years. Because these tubes or pipes are mass produced in various sectors, the industries need increased productivity. It is now necessary to enhance orbital welding in order to achieve this.

4. Conclusion

This study revealed that welding is a dangerous job, but not all welders are aware of all of the risks. This is made worse for people who are not involved in welding activity and are in the vicinity of the welding area. In this regard, some welders and others are harmed by welding risks just because they are unaware of the dangers. This document suggests some safety procedures to adopt during welding operations to minimise arc welding dangers. Basic research on the microstructure of the weld metal, which has been ongoing since the beginning, technical progress by welding methods that are being promoted from the standpoint of fatigue and fracture of welds, and recent quality control technologies for welds and joining technologies were also discussed. These technologies have already reached the point of practical application, and their existence as significant key technologies in responding to the stringent criteria put on steel materials is vital and indispensible.

References

[1] K.Y. Benyounis, A.G. Olabi (2008); "Optimization of different welding processes using statistical and numerical approaches – A reference guide." *Advances in Engineering Software* 39 (2008) 483–496

[2] Ugur r Elme (2008) ; "Application of Taguchi method for the optimization of resistance spot welding process." *The Arabian Journal for Science and Engineering*, Volume 34, Number 2B

[3] H.J. Park, D.C. Kim, M.J. Kang, S. Rhee (2004); "Optimisation of the wire feed rate during pulse MIG welding of Al sheets." *Journal of Achievements in Materials and Manufacturing Engineering* Volume 27, by International OCSCO world

[4] Taguchi G, Konishi S (1987). Taguchi methods, orthogonal arrays and linear graphs, tools for quality American supplier institute. *American Supplier Institute*; 1987 [p. 8–35]

[5] Taguchi G, Konishi S (1987). Taguchi methods, orthogonal arrays and linear graphs, tools for quality American supplier institute. *American Supplier Institute*; 1987 [p.8–35].

[6] Fiebai, B. Awoyesuku (2011), E. A. Ocular injuries among industrial welders in Port Harcourt, Nigeria. *Clinical Ophthalmology*. 5, 2011, pp. 1261-1263.

[7] Dixon, A. J. and Dixon, B. F (2004). Ultraviolet radiation from welding and possible risk of skin and ocular malignancy. *The Medical Journal of Australia*. 181(3), 2004, pp. 155-157.

[8] Pires, I. & Quintino, L. Amaral, V. and Rosado, T (2010). Reduction of fume and gas emissions using innovative gas metal arc welding variants. *International Journal of Advanced Manufacturing Technology*. 50, 2010, pp. 557-567.

[9] Parmar, R. S. Welding Processes and Technology (2007). Second Edition. Delhi: *Khan Publishers*, 2007, pp. 760.

[10] Budhathoki, S. S. Singh, S. B. Sagtani, R. A. Niraula, S. R. Pokharel, P. K (2007). Awareness of occupational hazards and use of safety measures among welders: a cross-sectional study from eastern Nepal. *http://bmjopen.bmj.com/content/4/ 6/e004646.short*.

[11] Amza, G. Rontescu, C. Cicic, D. – T. Apostolescu, Z. Picǎ, D (2010). *Research on Environmental Pollution When Using Shielded Metal Arc Welding (SMAW)*. 72(3), 2010, pp. 73-88.

[12] Kadinda, D. J (2007). Assessment of Arc Welding Hazards to Welders and Residents Surrounding Welding Workshops in Tanzania- Case Study of Dar es Salaam. *Senior Project Report. Bachelor of Mechanical Engineering*. 2007, pp. 65.

[13] Khanna O P (2000), *Welding Technology*, Dhanpat Rai Publications.

[14] Kishore K, Gopal Krishna P V, Veladri K and Sayed Qasim Ali (2010), "Analysis of Defects Gas Shielded Arc Welding of AISI 1040

Steel Using Tagauchi Method", *ARPN Journal of Engineering and Applied Science*, Vol. 5, No. 1.
[15] Lenin N, Shiva Kumar M and Vigneshkumar D (2010), "Process Parameter Optimization in Arc Welding of Dissimiliar Metals", *ThammasatInt. J. Sc. Tech.*, Vol. 15, No. 3.

Chapter 10

DYNAMIC SPECTRUM SHARING USING COGNITIVE RADIO NETWORKS IN 5G TECHNOLOGY

Ramalakshmi. R.
Department of Electronics and Communication Engineering,
Ramco Institute of Technology, Rajapalayam, Virudhunagar District, TamilNadu, India

Abstract: The explosive quality of small-cell and internet of everything devices have hugely increased traffic hundreds. This increase has revolutionized the present network into 5G technology that demands increased capability, high rate and ultra-low latency. However, the deficiency of the spectrum resource creates a heavy challenge in achieving an economical management theme, this work aims to conduct in-depth survey on recent spectrum sharing (SS) technologies towards 5G development and up to date 5G-enabling technologies. The surveys and studies area unit classified into one in all the most SS techniques on the idea of specification, spectrum allocation behavior and spectrum access technique. Moreover, an in depth survey on cognitive radio (CR) technology in SS associated with 5G implementation is performed. Spectrum holes area unit the temporary space-time-frequency that's not in use by any commissioned or unlicensed user and changes over time. The distort model additionally needs data on the user's activity from the spectrum. Spectrum occupancy is sporadically monitored and detected by the system. Negligible interference happens once the system communicates over obtainable spectrum holes.

1. INTRODUCTION

1.1. 5G Technology

Mobile traffic is foreseen to extend a thousand times over subsequent decade, because of the proliferation of connected devices. as an example, the amount of connected devices had reached fifty billion in 2020, as well as good phones, connected vehicles, and Internet of Things (IoT). Additionally to such a 1000× information challenge, a good vary of applications area unit rising with totally different service needs, like increased reality, e-health, and e-banking.

Consequently, next generation (5G) wireless network must improve the network capability considerably to support huge mobile traffic, and in the meantime satisfy distinct service needs from numerous applications.

5G network must boost the capability considerably to accommodate the mobile traffic surge and various services. Initially, a lot of spectrum is needed. Because the commissioned spectrum is sort of restricted, cellular networks area unit currently increasing to utilize the unlicensed bands (e.g., LTE-Unlicensed or LTE-U), like 5GHz and 60GHz. Excessive efforts from each trade and academe are created to alter LTE to operation in 5GHz bands. Besides, the under-utilized spectrum from alternative systems like TV white house (TVWS) are often harvested and reused in cellular systems to extend network capability opportunistically, through advanced psychological feature radio technologies.

1.2 Challenges and Contributions in 5G Technology

The network can alter a heterogeneous spectrum pool in terms of convenience, bandwidth, etc. Secondly, abstraction spectrum employ ought to be effectively improved across the network by deploying various tiny base stations (SBSs). However, densely deployed SBSs will suffer from severe inter-cell interference, which can degrade each the network capability and user expertise.

To effectively improve the network capability and satisfy users' needs, economical spectrum sharing among cells plays an important role. However, spectrum sharing in 5G networks chiefly faces the subsequent challenges:
- Since {different totally different completely different} cells have different traffic hundreds with various applications, spectrum sharing ought to satisfy the differentiated quality of service (QoS) needs with heterogeneous and dynamic resources;
- Inter-cell spectrum sharing and intra-cell user planning ought to be collectively optimized, that are including power allocation. Existing works on spectrum sharing area unit largely supported central optimization that sometimes incurs extraordinarily high complexness, or solely give suboptimal solutions.

The main contributions area unit summarized as follows:

- A complete list of the foremost vital 5G-enabling technologies and a quick summary of every technology area unit provided.
- Different SS techniques area unit classified.
- Related SS surveys conducted from 2014 to 2019 area unit reviewed. The main target and contributions of every study area unit summarized and bestowed in an exceedingly table.
- Related studies on spectrum sharing techniques relevant to 5G networks area unit reviewed. The studies area unit classified into specification, spectrum allocation behavior and spectrum access technique that is additionally one in all the most spectrum sharing techniques.
- Related spectrum sharing works specializing in energy economical (EE) enhancements are mentioned.
- Cognitive Radio technology in spectrum sharing and alternative applications associated with 5G implementation area unit reviewed. Psychological feature Radio in spectrum sharing are often thought-about a possible technology that may propel 5G networks into the long run.

1.3 5G Enabling Technologies

Currently, the emergence of assorted devices victimization wireless and IoT technologies is increasing remarkably. Networks that mix totally different cell varieties and access technologies area unit known as HetNets. Providing the network is flexible [2], more analysis on implementing this network is vital to avoid interference and fulfil the standard of Service promise to finish users. Many challenges that area unit crucial for effective preparation, like problems in network designing, traffic management and radio resource management, were reported. For ideal large-data streaming in 5G HetNets, Associate in nursing improved version of ancient traffic prediction was planned.

Another study planned the event of a linear predictor that uses compressed sensing by adopting support vector classification. The predictor contains a easy structure, and its results area unit promising. The predictor's performance is best than that of the standard load prediction methodology. With relevance HetNet network designing, conferred a quick relinquishing technique that uses a wireless link signature supported the user location because the relinquishing authentication information. The techniques area unit time-varying, unpredictable and secured with physical coding to ensure a definite and safe relinquishing.

1.4 Large MIMO

There has been active analysis worldwide to develop the next-generation, i.e., fifth-generation (5G), wireless network. [The 5G network is predicted to support a considerably great deal of mobile information traffic and an enormous variety of wireless connections and attain higher value and energy-efficiency in addition as quality of service (QoS) in terms of communication delay, responsibility, and security. to the present finish, the 5G wireless network ought to exploit the potential of latest developments, as well as super dense and heterogeneous preparation of cells and big antenna arrays [i.e., large multiple-input, multiple-output (MIMO) technologies] and utilization of upper frequencies, significantly millimetre-wave (mm Wave) frequencies[5]. The potential advantages and challenges of the 5G wireless heterogeneous network (HetNet) that comes with large MIMO and metric linear unit Wave technologies.

1.5 Immoderate Lean style

Future 5G technology to modify an ultra-lean style, wherever 'always-on' signals area unit reduced to a clean minimum to realize Associate in nursing engineering network at a coffee operational value. The implementation of this ultra-lean design [3] will scale back network transmissions while not touching user information delivery. This feature is important for terribly dense native area units to scale back the general interference level for end-user performance at low-to-medium masses and is crucial to high-frequency bands wherever networks are nonetheless to be deployed. Ultra-lean style additionally desires special attention in terms of backward compatibility for low-frequency bands as a result of an outsized variety of terminals area unit already deployed. Moreover, the implementation of ultra-lean style in immoderate Dense Network (UDN) has exhibited important improvement in enhancing quality support, increasing output and saving energy as by experimentation confirmed by, WHO provided future analysis insights into the preparation of 5G.

154

2 Dynamic Spectrum Sharing

2.1 Spectrum Sharing and adaptability

Spectrum allocation for 5G is categorised into 3 main bands, namely, low, high and really high. The spectrum at frequencies below one gigacycle per second, significantly at 700 megacycle per second allows 5G coverage in wide areas and deep indoor coverage.

The spectrum at high frequencies with comparatively massive bandwidths below six gigacycle per second (at 3.4 GHz to 3.8 GHz) provides the required capability to support varied connected devices and guarantee high speed for at the same time connected devices. This spectrum delivers the simplest compromise between capability and coverage. At terribly high frequencies higher than twenty four gigacycle per second (e.g. 24.25 gigacycle per second to twenty seven.5 GHz) with terribly massive bandwidths, the spectrum provides ultra-high capability and really low latency. The cells at these frequencies have alittle coverage (from fifty m to two hundred m). The build-out of 5G networks in millimetre Wave bands can at first be targeted on areas with high traffic demand or specific locations or premises requiring services with extraordinarily high knowledge rates (in Gbps). This 'pioneer' millimetre Wave band conjointly provides ultra-high capability for innovative new services, so sanctioning new business models and sectors of the economy to profit from 5G[3].

The C-band of the spectrum, that ranges from 3300 megacycle per second to 5000 megacycle per second, is selected because the primary band to introduce 5G within the year 2020. The channel information measure provided for 5G should be a minimum of a hundred megacycle per second per network to satisfy the wants. The implementation is incredibly efficient as a result of the data rate are often increased while not network compaction prices. 5G's use of the C-band are often accomplished by the adoption of large MIMO for its acceptable complexness and capability to spice up peak, average and cell edge outturn.

The low frequency used for mobile also can be exploited by combining the 3300 MH to 3800 megacycle per second frequency together of the 5G options in 3GPP standards through the utilization of the co-existing LTE/NR transmission. Amongst all 5G-enabling technologies conferred, this work focuses on the potential of totally utilising the restricted spectrum band by reviewing recent SS studies dedicated to 5G technology. the rationale for such focus is that the promising answer for authorised shared spectrum. the most benefits square measure enhancements in spectrum exercise and increment in capability consistent with the various kinds of services needed.

Category	Spectrum	Coverage
Low Frequencies	< 2 GHz	• Wide areas • Deep indoor coverage
High Frequencies	2 to 6 GHz	• Focused areas • Relatively large bandwidths • Very high number of connected devices • High speed of concurrent connected devices
Very High Frequencies	> 24 GHz	• Small coverage areas (50 to 200 m) • High traffic demand • Very large bandwidths • Ultra-high capacity • Peak data rates (Gbps) • Very low latency

Figure 1. 5G Frequency Spectrum

2.2 Dynamic Spectrum Sharing in 5G Technology

Dynamic Spectrum Sharing (DSS) could be a new antenna technology that for the primary time permits the parallel use of LTE and 5G within the same band. The technology determines the demand for 5G and LTE in time period. The network then divides the out there information measure severally and decides dynamically that mobile communications customary it ideally uses the out there frequencies.

For the user, Dynamic Spectrum Sharing means: If you surf with a 5G smartphone among the radius of Associate in nursing antenna equipped with the technology, you're surf boarding within the 5G customary. On the opposite hand, if you surf with a 4G phone among the signal vary of a similar antenna, you surf with 4G. In short: one antenna, 2 networks.

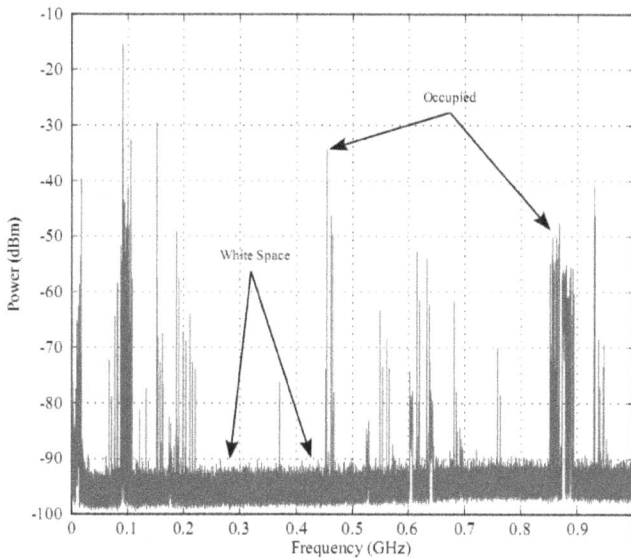

Figure 2. Used and Unused Spaces in Spectrum

2.3 Features

Not all service suppliers own spectrum licenses at intervals a Time Division Duplex (TDD) band to require advantage of 5G with its optimized quality of service, and to additional address the new market verticals (e.g., automotive and industrial), a network operator should transition to standalone (SA) mode, during which the 5G RAN is connected to the 5G core network[5]. There square measure many intermediate steps resulting in a standalone preparation, and every operator will follow the trail outlined by its 5G strategy.

Due to the occupation of their Frequency Division Duplex (FDD) - based spectrum assets, service supplier's square measure forced to decide on between 2 pricey options:

- Acquiring new spectrum
- Reframing spectrum already in use

However, the 5G NR normal offers the chance of adapting to existing LTE deployments and sharing the spectrum used completely by LTE

157

these days. The sanctionative mechanism is Dynamic Spectrum Sharing(DSS), that permits 5G NR and 4G LTE to exist whereas victimization a similar spectrum. Within the future, DSS permits network operators to produce a coverage layer for 5G victimization the lower frequency bands. Some network operators already make the most of DSS, and an outsized scale preparation is predicted late 2020 to early 2021.

2.4 Testing of DSS

In depth testing is needed for DSS implementation. This includes lab-based LTE and 5G user instrumentality testing also as network performance measurements victimization scanners (sensitive receivers) and devices to estimate coverage and end-to-end (E2E) performance.

The activation of DSS at intervals the network shouldn't produce any interference for the prevailing LTE preparation. LTE-only devices should not suffer any impact once configuring MBSFN sub frames. With MBSFN active, E2E output tests square measure needed to confirm bottom impact on LTE performance. Whereas 5G NR, as well as SSB, is transmitted at intervals MBSFN sub frames, receiver sensitivity tests for LTE devices should be favorable, to confirm necessities square measure still met by the device once 5G NR is gift.

A 5G NR capable device should be ready to synchronize in time and frequency with the 5G RAN once SSBs square measure transmitted at intervals MBSFN designed sub frames. once 5G NR is shipped in non-MBSFN sub frames victimization Associate in Nursing LTE CRS rate-matching pattern for NR's PDSCH, a knowledge output check is up to verify correct implementation of the stack features[7]. Advanced device testing includes dynamic planning procedures that mimic the E-UTRA NR resource coordination procedure.

2.5 Operation of DSS with 5G

With simply 3 sub frames offered for 5G NR, the technology operates underneath its potential. DSS in addition allows the employment of sub frames that area unit dedicated to LTE and not designed for MBSFN via 2 distinct features:

- Depending on the MIMO mode, commonplace LTE sub frames embody cell-specific reference signals (CRS) mapped to bound resource parts within the time-frequency grid. AN LTE terminal uses CRS for channel estimation and to take care of full synchronization in time and frequency [3]. To change NR to use these sub frames, rate-matching round the LTE CRS has been adopted.
- A different extra position for the mapping of the physical knowledge shared channel (PDSCH) reception reference signal (DMRS) is supported, once more to avoid collision with LTE CRS. This feature could be a device capability; the device signals its support of this practicality to the network throughout the initial registration method.

3. Cognitive Radio Networks

3.1 Cognitive Radio Network Cycle

A Cognitive radio could be an absolutely reconfigurable device which might observe and alter or adapt its communication parameters for facultative secondary usage of the spectrum and yield an economical usage of the spectrum. However, it's necessary to form positive that the employment of the spectrum by the secondary user doesn't cause any interference to the legitimate users. Thus, metal is wide considered one in every of the foremost promising technologies for future wireless communications. to form radios and wireless networks really cognitive, however, is by no suggests that a straightforward task, and it needs cooperative effort from varied analysis communities, together with communications theory, networking engineering, signal process, theory of games, software-hardware joint style, and reconfigurable antenna and radiofrequency style.

The availability of the unused spectrum may be determined by the subsequent techniques:
- Passively sensing the spectrum.
- Using separate transmitters to point the spectrum convenience
- Using location data of the radio to visualize the information or frequency spectrum usage.

159

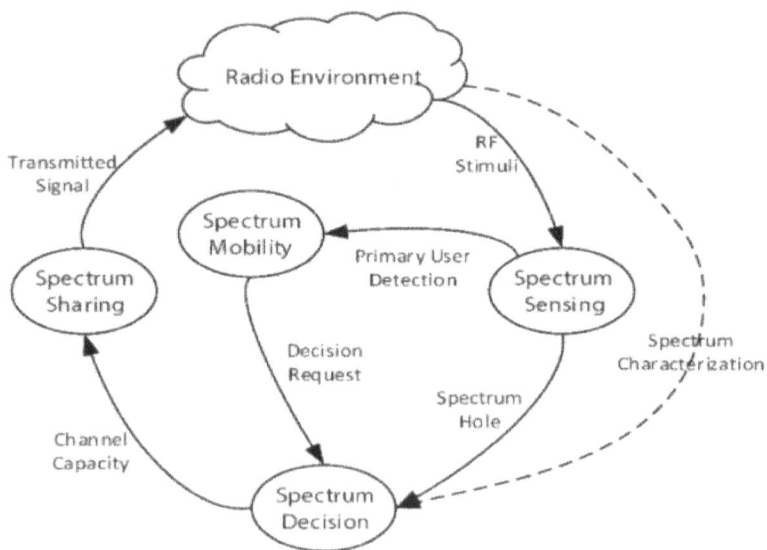

Figure 3. Cognitive Radio Cycle

3.2 Cognitive Radio design

A Cognitive radio network consists of primary networks similarly as secondary networks. A primary network contains of 1 or additional PUs and one or additional primary base stations. The PUs square measure authorized to use the spectrum and square measure coordinated by the first base stations. PUs communicate among one another through the bottom station solely. Generally, the PUs similarly because the primary base stations don't have CR properties.

On the opposite hand, a secondary network contains of 1 or additional SU and should or might not contain a secondary base station. For SUs, the spectrum access is managed and handled by the secondary base station that acts as a hub/access purpose for the SU network. The SU below the vary of constant base station communicate with one another through the bottom station. If quite one secondary base station shares one spectrum band then their spectrum usage and coordination is finished by a central spectrum broker. A group of SU can even connect with one another and communicate among themselves while not the presence of the secondary base station.

As the SU mustn't cause interference with the PUs transmissions, all the SU at the side of the secondary base stations square measure equipped with the CR properties. Thus whenever SU find the presence of a PU during a spectrum band they ought to in real time stop victimisation that band and may move to another obtainable band to avoid interference with the PU transmission.

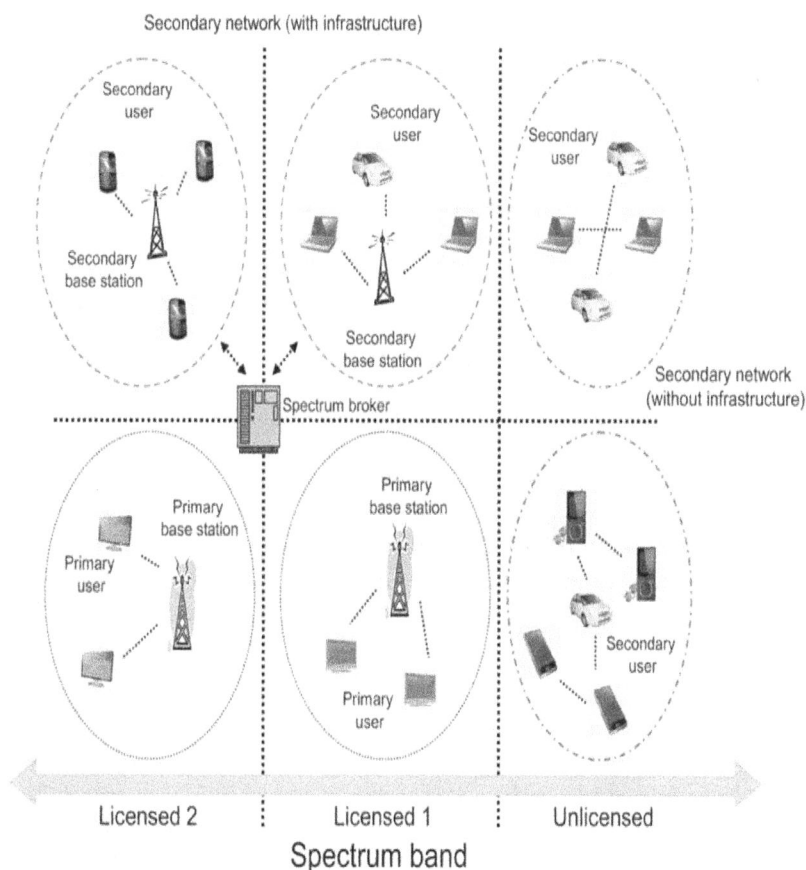

Figure 4. Cognitive Radio Architecture

161

3.3 Characteristics of Cognitive Radio Networks

Some cognitive radio network characteristics are as follows:

- Operating atmosphere sensing: CRs operate in an exceedingly multi-dimensional atmosphere which may embody cooperative or non-cooperative emitters that may toggle on and off, adapting to native changes further as traffic masses that vary speedily. so as to perform its task properly, a CR should amendment in accordance to the ever-changing atmosphere and it ought to be ready to inform alternative devices within the network relating to the modified configuration.

- Operational state languages: Operational state languages square measure used for data sharing in an exceedingly chromium network. As mentioned higher than chromium ought to inform its states and observations to alternative nodes in its network. The language that CRs use for this purpose is named operational state language. The data that a chromium sends could be a listing of all emitters that it recently detected.

- Distributed Resource Management: The spectrum may be a distributed resource. Therefore, use of a spectrum band at one location makes it untouchable elsewhere. Therefore, the allocation of spectrum resources should be wiped out a balanced fashion. Numerous algorithms are developed to handle the allocation and managing the distributed spectrum and resources supported traffic masses.

3.4 Functions of psychological feature Radio Network

Spectrum Sensing: To avoid interference, the spectrum holes (bands not being employed by the PUs) ought to be detected. Pu detection technique is that the best approach during this respect. The spectrum sensing techniques square measure primarily divided into 3 classes, that square measure transmitter detection, co-operative detection and interference based mostly detection.

Spectrum Management: There is a necessity to capture the simplest out there spectrum to satisfy the user communication needs. CRs ought to choose the simplest spectrum band to satisfy quality of service needs over all spectrum bands. The management operate is classed as spectrographic analysis and spectrum detection.

Spectrum Mobility: It is the method wherever a chromium user exchanges the frequency of operation. They aim to use the spectrum in an exceedingly dynamic manner by permitting the radio terminals to work within the best out there waveband. The shift to a stronger spectrum should be seamless.

Spectrum Sharing: It is of utmost importance to supply a good spectrum planning policy. It is additionally one in every of the foremost necessary challenges in open spectrum usage. Within the existing systems it corresponds to the prevailing waterproof issues.

3.5 Different scenarios in Cognitive Radio

There are two different types of spectrum sharing scenarios. They are,

- o Cooperative scenario
- o Non-cooperative scenario

In cooperative scenario, a primary user provides secondary users with all data relating to the occupancy of the spectrum and concerning the unused spectrum in order that the secondary users create use of that unused spectrum and prevent from the occupied spectrum.

Within the non-cooperative scenario, a secondary user must sense the spectrum for the unused spectrum and use that spectrum band while not inflicting any interference to the first user.

Within the cooperative scenario, a malicious user will masquerade because the primary user and supply false data to the secondary user relating to the occupancy of the spectrum, like the spectrum is unoccupied and therefore the secondary user will use the first user occupies the spectrum. With the data provided, the secondary user tries to occupy the spectrum and as a result, interference takes place between the first user and secondary user.

163

3.6 Functions of Cognitive Radio Network

There are three different classes of Cognitive radio paradigms.

- Underlay
- Overlay
- Interweave

The Underlay cognitive radio paradigm is employed once the interference between the psychological feature users and non-cognitive users is below a precise threshold.

In Overlay cognitive radio paradigm communication is provided by subtle signal process.

The twine cognitive radio paradigms opportunistically exploit the white areas while not inflicting any interference to the opposite transmissions. Generally, twine cognitive radio system is employed.

3.7 Security in Cognitive Radio Network

Cognitive Radio (CR) offers nice potentials for reconciling networks and dynamic spectrum access for increased spectrum utilization. However, there square measure security vulnerabilities in CR networks. Within the physical layer, a secondary or CR node initial senses the channel atmosphere to work out spectrum holes that is subject to attack with associate in nursing assailant manipulating the network atmosphere.

In access behavior or a medium access management (MAC) layer, acts reuse CR, inconsiderate CR or cheating CR may occur thanks to the versatile or open access mechanisms in CR networks.

3.8 Benefits of Cognitive Radio Network

- Senses the oftenness atmosphere for the presence of white areas.
- Manages the unused spectrum.
- Increases the potency of the spectrum utilization considerably.

- Use the unused spectrum for brand new business proposi-
  tions, like providing high speed web within the rural areas
  and high rate network applications like video conferencing
  will be created.

4. Working

The flowchart indicates the working principle used in spectrum
sharing.

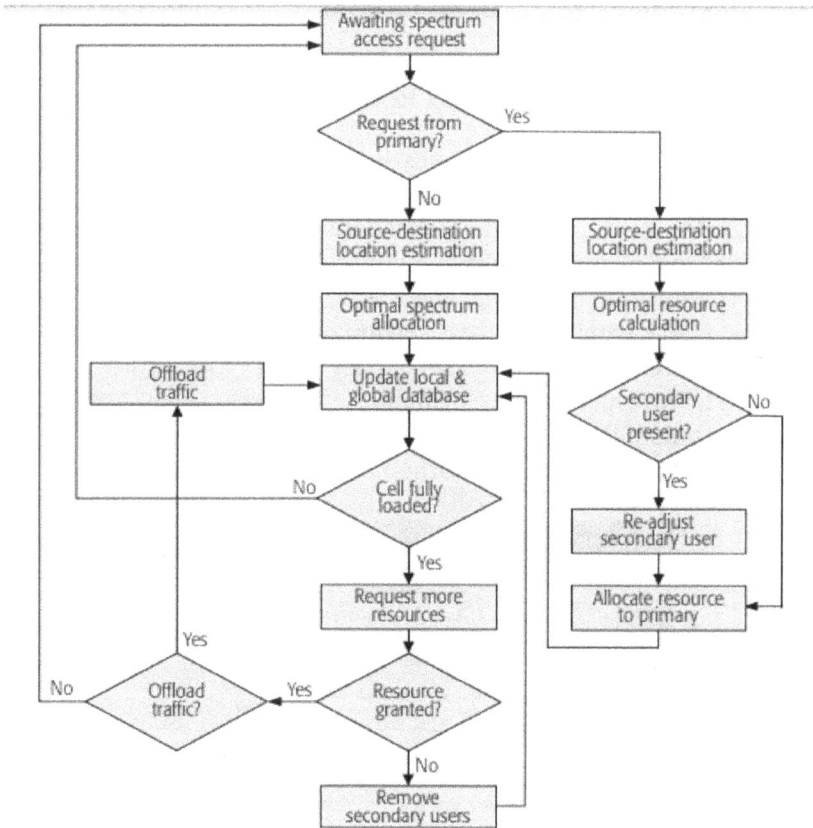

Figure 5. Resource Management process at the BS

The software used is Matlab. The version of Matlab Software is
2019b.The following are the steps used in Matlab software to find
the Sharing of Spectrum.

Step 1: Awaiting for spectrum access request
Step 2: Request from primary user

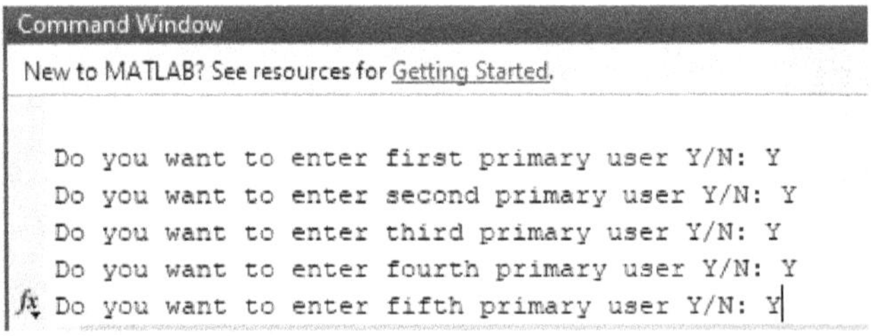

```
Command Window
New to MATLAB? See resources for Getting Started.

    Do you want to enter first primary user Y/N: Y
    Do you want to enter second primary user Y/N: Y
    Do you want to enter third primary user Y/N: Y
    Do you want to enter fourth primary user Y/N: Y
fx  Do you want to enter fifth primary user Y/N: Y
```

Figure 6. Allowance of Primary users

Step 3: If "No", estimate source-destination location.
Step 3.1: Allocation of optimal spectrum
Step 3.2: Update the local and global database
Step 4: If "Yes", estimate source-destination location.
Step 4.1: Allocation of optimal spectrum
Step 4.2: Check whether there is secondary user present
Step 4.3: If "Yes", re-adjust the secondary users
Step 4.4: If "No", allocate it to primary
Step 4.5: Update the local and global database
Step 5: If the cell is fully loaded, then request more resources

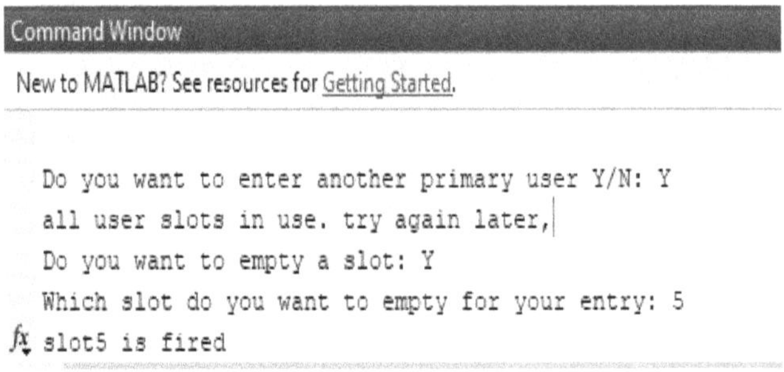

```
Command Window
New to MATLAB? See resources for Getting Started.

    Do you want to enter another primary user Y/N: Y
    all user slots in use. try again later,
    Do you want to empty a slot: Y
    Which slot do you want to empty for your entry: 5
fx  slot5 is fired
```

Figure 7. Allowance of Additional User

Step 6: If the request is not guaranteed, then remove the
Secondary users.
Step 7: If the request is guaranteed, then check for the traffic
Step 7.1: If "Yes", Remove users
Step 7.2: If "No", then repeat from step 1

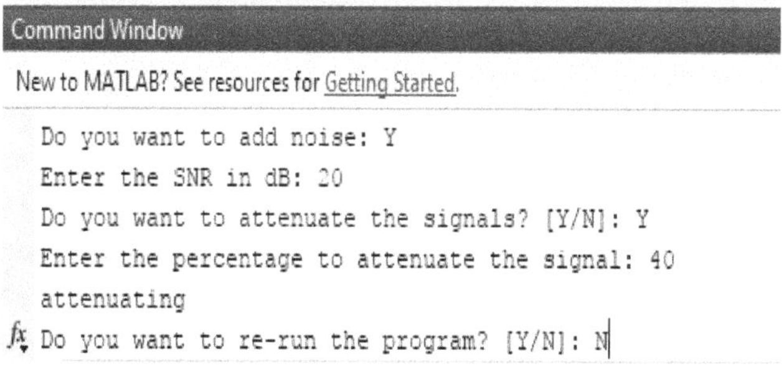

```
Command Window
New to MATLAB? See resources for Getting Started.
  Do you want to add noise: Y
  Enter the SNR in dB: 20
  Do you want to attenuate the signals? [Y/N]: Y
  Enter the percentage to attenuate the signal: 40
  attenuating
fx Do you want to re-run the program? [Y/N]: N
```

Figure 8. Allowance of Noise and Attenuating the Signal

5. Results and Discussion

The proposed method focuses on the development of spectrum sharing without any interference.When the code gets executed,it checks the spectrum which slots are in use and which are not in use.The slots which are not in use gets occupied by the secondary users and the primary users using the slots may get a chance to fired out from the slot also.There will no interference between the users sharing the spectrum at the same time and there will be low latency,high throughput.

The major goal is to minimise delay whilst sharing a spectrum and efficient deployment of 5G NR networks inunpaired assignments andthe alignment of uplink/downlink transmissions for slot and frame synchronisation. To minimise delay whilst sharing a spectrum and efficient deployment of 5G NR networks in unpaired assignments and the alignment of uplink/downlink transmissions for slot and frame synchronisation.

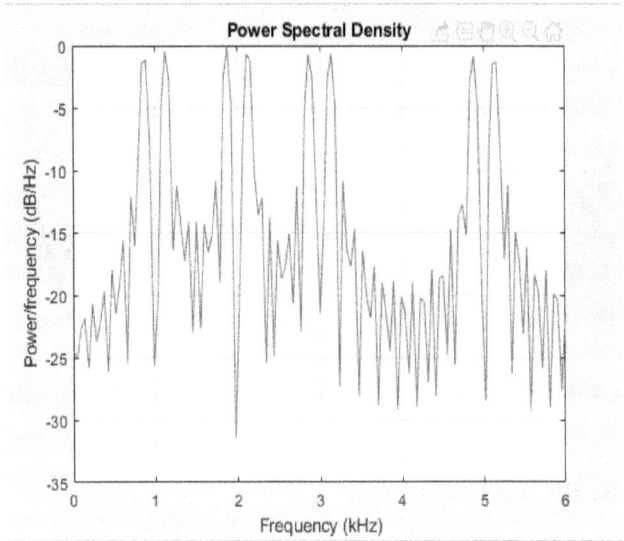

Figure 9. Normalization by Spetcral Resolution

The power spectral density is obtained in figure 9. The spectrum is shared among all the users and is efficient. It allows all the primary users to share the spectrum.

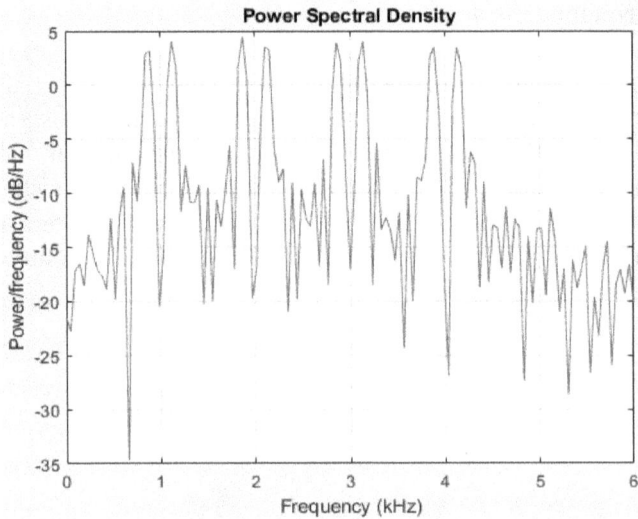

Figure 10. Power Variations when noise is added

168

The power spectral density when noise is added is obtained in figure 10. The spectrum is shared among all the users and is efficient. It allows all the primary users and secondary users to share the spectrum.

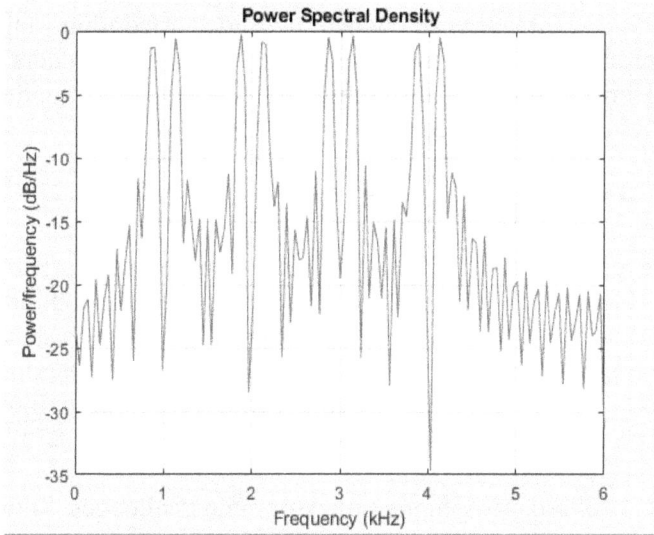

Figure 11. Power Variations when attenuated

The power spectral density is obtained in figure 11. The spectrum is shared among all the users and is efficient.

4. Conclusion

New dynamic spectrum access schemes are proposed for cognitive radio wireless networks without and with buffering for the cognitive SU to avoid direct blocking. Performance metrics for SU are developed with respect to blocking probability, interrupted probability, forced termination probability, non-completion probability and waiting time. The result indicates that the buffer is able to significantly reduce the SU blocking probability and no completion probability with very minor increased forced termination probability.

The main 5G development issues are related to EE and interference. Multi-user interference in dense networks is a serious issue, especially in licensed bands. Propagation losses and the interference that fluctuates in signal quality must be addressed at the hardware level. The current solution can reduce the interference but at the cost of energy consumption and infrastructure. Multi-hop communication increases spectrum efficiency but has a high switching delay. Interference management is also needed for cognitive NOMA to handle inter-network interference, which degrades reception reliability. This comprehensive survey is expected to aid researchers in addressing important issues in achieving successful 5G applications. The analytic model has been verified by the simulation.

5. Future Scope

In future, all the users will use the spectrum at a similar time with none problems. DSS allows the short and cost-efficient build out of sturdy 5G services. It's employed in broad coverage areas, exploitation existing spectrum in mid- and low-band frequencies. Operators with giant amounts of frequency spectrum would have an interest to deploy DSS to assert 5G capability.

Dynamic SS is receptive many policy-domain challenges. Knowledge base analysis is needed for spectrum usage and access to manage the sharing of restricted spectrum resources expeditiously within the fields of market- and non-market-based mechanisms. different open issues in SS embrace spectrum management and spectrum assignment/allocation, metrics to quantify spectrum usage, interference existence and management, security and social control, radio software system and hardware, protocols and standards, experimentation, testing and standardization and regulative, policy and economic problems. Interference happens once mammal genus begin their transmission while atomic number 94 transmission is current within the spectrum as a result of the dearth of sensing accuracy. This is often a significant issue in hybrid SS, particularly in systems that utilize authorized bands.

Moreover, propagation losses and interference occur from looking the hidden terminals of PUs and incomprehensible transmission opportunities. Software-defined analysis on SS style has conjointly been conducted to cut back device-level limitations in manufacturing inaccurate results and reducing fluctuation in signal quality. Another means that of assuaging the interference from the design perspective is to put stress on physical layer style so as to strengthen the spatial focusing impact from mm Wave and big MIMO. The mackintosh layer style can specialize in the way to utilize the focusing impact to accommodate additional users. This may cause different problems within the mackintosh layer, and re-investigation is needed once the spatial focusing impact is taken into account. The problems square measure admission management, relinquishing and security. Within the future, once applied science is improved by using energy harvest home technologies at secondary BSs, a difficulty may be encountered in terms of users' QoE because of the intermittent arrival of renewable energy. Therefore, the authors urged that the energy standing of secondary BSs should be thought of throughout SS.

References

1. Akhtar.A.M, X. Wang, and L. Hanzo (2016), 'Synergistic spectrum sharing in 5G HetNets: A harmonized SDN- enabled approach', *IEEE Communication Magazine*, vol. 54, no. 1, pp. 40 - 47.
2. Anandakumar.H and K. Umamaheswari (2019), 'Cooperative spectrum handoversin cognitive radio networks in Cognitive Radio, *Mobile Com-munications and Wireless Networks'*, Cham, Switzerland: Springer, pp. 47 - 63.
3. Bairagi.K et.al. (2018), 'QoE enabled unlicensed spectrum sharing in 5G: A game-theoretic approach', *IEEE Access*, vol. 6, pp.50538 - 50554.
4. Beltrán.F and M. Massaro (2018), 'Spectrum management for 5G: Assignment methods for spectrum sharing', in *Proc. 29th Eur. Regional ITS Conference*.
5. Dludla.G, L. Mfupe, and F. Mekuria (2018), 'Overview of spectrum sharing models: A path towards 5G spectrum toolboxes in Information and Communication Technology for Development' for Africa. Cham, Switzerland: *Springer*, pp. 308 - 319
6. Jiang.C, B. Wang, et.al. (2017), 'Exploring spatial focusing effect for spectrum sharing and network association', *IEEE Transition Wireless Communication*, vol. 16, no. 7, pp. 4216 - 4231.
7. Mishra.S, S. S. Singh, and B. S. P. Mishra (2019), 'A comparative analysis of centralized and distributed spectrum

sharing techniques in cognitive radio Computational Intelligence in Sensor Networks'. Berlin, Germany: *Springer*, pp. 455-472.

8. Mustonen.M (2017), 'Analysis of recent spectrum sharing concepts in policy making: Dissertation', Ph.D. dissertation, Univ. Oulu, Oulu, Finland.

9. Rattaro.C et.al (2018), 'Multi-resource allocation: Analysis of a paid spectrum sharing approach based on _uid models', *IEEE Transition Cognitive Communication Network*, vol. 4, no. 3, pp. 607 - 617.

10. Wang.H et.al. (2017), 'Negotiable auction based on mixed graph: A novel spectrum sharing framework', *IEEE Transaction Cognitive Communication Network*, Vol. 3, No. 3, pp. 390 - 403.

11. Yang.C et.al. (2016) 'Advanced spectrum sharing in 5G cognitive heterogeneous networks', *IEEE Wireless Communication*, vol. 23, no. 2, pp. 94 -101.

12. Zhang.N, S. Zhang et.al. (2017), 'QoE driven decentralized spectrum sharing in 5G networks: Potential game approach', *IEEE Transaction Vehicle Technology*, vol. 66, no. 9, pp. 7797-7808.

Chapter 11

Research Issues and Challenges of Wireless Networks

R. Prabhu and R. Guruprasath
Department of Electronics and Communication Engineering,
Gnanamani College of Technology, Tamilnadu, India

G. Srividhya
Department of Electrical and Electronics Engineering,
Gnanamani College of Technology, Tamilnadu, India.

Abstract

The necessity for wired and wireless networking has grown more important as civilization progresses. According to security, each of these methods of networking offers benefits and downsides. Wired networking has various hardware requirements, as well as a wider variety of features and advantages. Wireless networking considers the range, mobility, and various sorts of hardware components that are required to set up a wireless network. As you read on, you'll learn about various sorts of network setups and the security precautions that must be followed to guarantee a safe network.

1. Introduction

The phrase "wireless communication" was coined in the nineteenth century, and wireless communication technology has progressed in the years thereafter. It is one of the most essential means of information transfer from one device to another. Using electromagnetic waves such as IR, RF, satellite, and others, information may be conveyed via the air without the need of cables, wires, or other electrical conductors. Wireless communication technology nowadays refers to a wide range of wireless communication devices and technologies, including smart phones, computers, tablets, laptops, Bluetooth technology, and printers. This page provides an overview of wireless communication and its many variants. The phrase "wireless" refers to the transfer of data over a long distance without the need of wires, cables, or other electrical conductors. Wireless communication is an essential mode of data or information transfer to other devices. In a wireless communication technology network, communication is established and information is conveyed in the air, without the need of wires, using electromagnetic waves such as radio frequencies, infrared, satellite, and so on. The basic purpose of Bluetooth technology is to enable you to connect a range of different electronic gadgets wirelessly to a system for data transmission and exchange. Bluetooth connects cell phones to hands-free earpieces, wireless keyboards, mice, and microphones, as well as wireless keyboards, mice, and microphones, to laptops. Bluetooth technology serves a variety of purposes, although it is most widely employed in wireless communication. There are wireless broadband systems that provide rapid Web browsing without the need for a cable or DSL connection (Example of wireless broadband is WiMAX). Despite the fact that WiMAX has the capacity to transmit data speeds of more than 30 Megabits per second, carriers often offer data rates of 6 Mbps or less, making the service much slower than traditional broadband. The real cost of WiMAX data varies greatly depending on the distance from the transmitter. Sprint's 4G technology uses WiMAX, which is one of the variants of 4G wireless accessible in phones.

Satellite communication is a self-contained wireless communication technology that is extensively used across the globe to enable users to remain connected from practically anywhere on the planet. When a signal (a modulated microwave beam) approaches the satellite, the satellite amplifies the signal and sends it back to the antenna receiver on the earth's surface. The space segment and the ground segment are the two basic components of satellite communication. The ground section is made up of permanent or mobile transmission, reception, and supporting equipment, whereas the space segment is mostly made up of the satellite itself. Wireless infrared communication IR radiation is used to transfer information inside a device or system. IR refers to electromagnetic radiation with a wavelength longer than red light. It's used for security, TV remote control, and short-range communications, among other things. IR radiation is found between microwaves and visible light on the electromagnetic spectrum. As a result, they may be employed as a communication tool.

A photo LED transmitter and a photo diode receptor are necessary for effective infrared communication. The IR signal is sent by the

LED transmitter as non-visible light, which is caught and recorded by the photoreceptor. As a result, information is exchanged between the source and the destination in this manner. Mobile phones, TVs, security systems, computers, and other devices that allow wireless communication may be used as the source and destination. Open radio communication was the first wireless communication technology to gain broad adoption, and it continues to serve a role today. Multichannel radios are portable and allow users to communicate over short distances, while citizen's band and marine radios provide communication for sailors. With their powerful broadcasting gear, ham radio enthusiasts exchange data and operate emergency communication aids during catastrophes, and can even convey digital information throughout the radio frequency spectrum. With the advancement of network and communication technologies, the difficulty of wire has been eliminated from people's lives, and WSN has a broad range of applications and applications in the areas of remote sensing, industrial automation control, and household appliances, among others. WSN has strong data collecting, transmission, and processing capabilities. When compared to other options, it offers a lot of benefits.

Traditional wired networks include advantages such as ease of organisation, minimal environmental impact, low power dissipation, and cheap cost. Near-field wireless communication technology, such as Bluetooth, wireless local area network (WLAN), and infrared, is now extensively employed. However, they have a lot of drawbacks, including as complexity, high power consumption, limited distance, and small-scale networking. As the times demand, a new form of wireless net technology-Zigbee arises to meet the requirement for low power consumption and low speed among wireless communication devices. We shall present Zigbee networking technology and applications in this article. How Zigbee and RFID may be utilised together in applications. First, Zigbee is discussed, followed by its benefits and uses, and ultimately, its fusion with RFID and applications.

With the ability to coordinate reciprocal communication among thousands of small sensors, this standard is a must-have. These sensors can communicate data from one sensor to another using radio waves at a low energy cost and great efficiency. When compared to other wireless communication technologies, ZigBee uses the least

175

amount of energy and is the least expensive. Because of its sluggish data rate and limited communication range, ZigBee technology is ideal for agricultural fields with minimal data volumes. This technology's technical characteristics also make it the ideal option for wireless sensor networks. As a result, it offers practical value when used in a crop environmental monitoring system. The following are the characteristics of ZigBee. ZigBee employs a range of power-saving modes to ensure that two AA batteries can run it for at least six months to two years. To minimise rivalry and conflict while transferring data, ZigBee employs the avoidance collision mechanism in CSMACA and pre-sets a certain time slot for a fixed bandwidth communications service. Each packet delivered by the receiver must wait for confirmation at the MAC layer, which uses a fully confirmed data transfer method. Zigbee has self-organizing features, which allow one node to detect other nodes without the need for human intervention and automatically connect them to form a complete network. It also has a self-recovery mechanism, which allows the network to repair itself when a node is added or removed, a node's location is modified, or a breakdown occurs. It may also alter the topological structure to guarantee that the system runs smoothly without the need for human interaction. Provide real-time collection, storage, monitoring, and reporting.

Its functions are separated into two primary components, processing equipment from a distant terminal node of information, and it may overflow the police at any moment, such as defining parameters for the production environment to accomplish efficient monitoring and management. Data Monitoring: to receive information from the ZigBee network and enter the corresponding data into the database; to receive instructions from the managers and command frame format in accordance with the configuration commands, GPRS module through the command issued to the ZigBee network and perform the action; to receive instructions from the managers and command frame format in accordance with the configuration commands; to receive instructions from the managers and command frame format in accordance with the configuration commands; to receive instructions from the managers and command frame format in accordance with the configuration Data Management: The database may be located, and data from the current ZigBee network information, such as: ambient temperature, pressure, overrun alert, and peak period, can be queried. Time for a ZigBee end-node to wake up from hibernation and begin data collecting, ZigBee routing node to issue a wake-up call from time to time. Send the message, and then go into hibernation.

2. Wireless Networks

The data delivered to the ZigBee coordinator node, gateway GPRS module, and data uploaded to the remote monitoring centre will be collected by ZigBee routing nodes. RFID is a non-contact automated identification system that detects targets and provides access to pertinent data using radio frequency signals. The identification process does not need human intervention and may be used in a range of challenging conditions. However, if there is no network to send data, it will be impossible to take benefit of the technology. The typical wired network may not be a superior technique to accomplish under the effect of environmental circumstances. The characteristic of a wireless sensor network is that it has no centre and may self-organize. It is a strong complement to RFID and can overcome the disadvantage of weak anti-interference. Short effective transmission distance Based on information-fusion technology's ZigBee and RFID technologies, the former is used to monitor target environment conditions, while the latter is used to identify target items. Complementary and interdependent technologies may effectively handle the issue of RFID data transmission in mines, as well as better perceive the safety hazards that exist in coal mines. In a few years, Zigbee wireless communication technology will be employed in the areas of industry control, industrial wireless location, home network, building automation, medical equipment control, mine safety, and so on. The primary application sectors will be home automation and industry control. In households, Zigbee wireless communication is used. The notion of smart house and home automation has been popular as people's lives have progressed, but it must relate to the transfer of information and signal if it is to be realised, which makes wiring wires difficult. Zigbee is a revolutionary short-range wireless communication technology that is specifically developed for low-speed and low-power wireless communication applications, making it suitable for building a family wireless network. Home temperature management, remote control of interior lighting systems, and automated curtain adjustment are all simple to implement.

In a metre reading system, Zigbee wireless communication technology is used, and the monitoring centre just has to interpret and

177

compute data collected from customers in order to determine their power use. After that, the monthly electric bill is withdrawn from users' energy accounts, and personnel who are required to read the metre at the user's house are prevented from reading the metre while users are not at home. It is more crucial to be utilised in safety than it is to work quickly for employees. Introduces an experimental Zigbee-based home security monitoring and alarming system capable of monitoring door and window magnetic contact, smoke, gas leaks, water flooding, providing simple controls such as turning off valves, and sending alarms to the residential area security network, among other things. In industries and businesses, Zigbee wireless communication technology is used. It is used in the information system of coal preparation firms to avoid all of the drawbacks of the old cable network system, and it greatly enhances the degree of information automation, automation, and management. In the ARM NC system network, Zigbee wireless communication technology is used. The enhanced approach can ensure the processing efficiency of the NC system with satisfactory accuracy and data transmission speed, according to the findings of the experiments. A unique Zigbee-based laser alarm system aimed at substation perimeter safety is suggested in. It comprises of a laser railing security subsystem and a data central monitoring subsystem. Zigbee wireless technology is used to communicate between the two subsystems, and a real-time human-machine interface may be given for Zigbee wireless communication in mine.

Zigbee technology is used in the Miner's Lamp Monitoring in order to improve production safety and employee safety. This system may provide underground staff orientation, as well as monitoring and control of the miner's lamp's charge condition, as well as very effective control and management of miner's lamp usage. Using the current subterranean network and extension Zigbee nodes, the system may more simply enhance humidity, gas, and other sensors, achieving mine environmental goals. The better approach has been explored in Zigbee has been extensively employed in various fields because to the advantages of low power consumption and cheap cost, and it is excellent for wide-scale application. However, there are some issues presently.

There are certain places where it is difficult for humans to change the batteries of nodes, or there is a relatively large number of nodes that is difficult to change presents an improved design, the coordinator just deal with the tasks on the Zigbee network, the rest tasks will be processed by anode. It's important to extend the life of the Zigbee network. The design of the Zigbee routing protocol has a significant aim. An energy-conscious routing system It is shown that EA-AODV can conserve energy and increase the performance of the Zigbee network. The physical layer, radio resource control (RRC) layer, and non-access stratum (NAS) layer are all part of the 5G network protocol stack. Each layer has its own protocols for carrying out its operations, such as the protocols for joining and removing devices from the network, as well as the protocols for paging devices that give alerts of incoming calls and text messages. As a consequence, in addition to the vulnerabilities in the inherited functionality from 3G and 4G networks, the new technologies and protocols introduced in 5G bring new attack surfaces that have yet to be thoroughly examined in terms of security and user privacy. Designing protocols that can meet their stated security and privacy assurances even in the face of adversaries, as well as the capacity to reason about stateful protocols that use cryptographic constructions, will be necessary for a strong 5G ecosystem.

3. Applications of Wireless Networks

While thorough analysis helps in identifying the core causes of vulnerabilities, effective mitigation strategies and secure solutions are also required to safeguard next-generation cellular networks from sophisticated attacks. Phony base stations, fake emergency warnings, identity exposure assaults, and some types of side-channel attacks are all protected by these safeguards. However, because of the structure of the cellular network ecosystem and the varied motives of its players, adopting security approaches for widespread usage in cellular networks is equally difficult. WSN can also be defined as a network of possibly low-size and low-complexity devices known as nodes that are capable of sensing the environment and communicating gathered information from the monitored area; the gathered data can be transmitted directly or via multiple hops to a sink, which can then use it locally or connect to other networks (e.g. the internet) through gateway nodes. The performance of topology control techniques in WSNs is affected by node deployment, which is application dependent. Deterministic or randomised deployments are also possible. Sensors are manually installed and data is transmitted via pre-determined pathways in deterministic deployment. Random

179

node deployment, on the other hand, scatters sensor nodes at random, resulting in an ad hoc architecture. Wireless sensor networks, unlike other networks, are designed to capture sensory data, such as water quality and pollution data for a specific section of a river. The wireless sensor network data query system offers users with a SQL-like sensing data query interface that is simple to use. Users may query sensor data in the same way they query standard relational database systems, considerably reducing the complexity of developing wireless sensor network applications[2]. Although conventional database query processing technology has shown positive research findings, these results were obtained mostly in a PC context and are not applicable to wireless sensor networks. Many researchers have looked at the characteristics of restricted sensor nodes, frequent network topology changes, and simple node failure in recent years. For sensor networks, a novel query processing technique is presented. Take, for example, the University of California at Berkeley's TinyDB system.

A TelosB-based sensor hardware platform is created to accommodate sensors such as temperature, humidity, and pressure, in order to address concerns found in present query processing research. A total of 50 sensor nodes are installed on three levels of a building to monitor the building's surroundings, conduct people placement, and scene perception in order to produce an image platform for experimentation The current data collecting query technique and the wireless sensor network data query system TinyDB were both evaluated on the experimental platform. The following phenomena were discovered as a result of the experiment: To begin with, despite the fact that the wireless communication link's bandwidth may potentially reach 256KB/S, the current query processing method delivers the query response to the base station at a pace of no more than 2 KB/S; Second, after the query processing algorithm has run for a period of time, some nodes in the network's sensing data cannot be returned to the base station; the phenomenon of phenomenon 1 occurs because the existing query processing algorithm assumes that the network is deployed in an unrestricted environment. In a constrained wireless sensor network, the sensor node has fewer data forwarding channels. In the most severe instance (such as a sensor node placed in a corridor), only one routing route provides the query response, causing the wireless communication channel to compete. When the intensity is increased, the likelihood of packet collision rises, and the throughput of query result transmission falls. Because there are cut sites in the network, symptom 2 arises. Node failures are common due to issues such as a bad deployment environment, software failure, and inadequate power, and the failure of the cut point leads the network to be disconnected.

4. Conclusion

Wireless sensor networks rely heavily on query processing. There are benefits in certain unique applications that regular technology do not offer. Although considerable progress has been achieved, there are still numerous issues that need more investigation and debate. The wireless sensor network is abstracted into a distributed database in this research in order to analyse and create an effective cut-point query technique for network reconfiguration and performance diagnostics. Simultaneously, a node data transmission and reception scheduling strategy for the limited deployment space is proposed in order to improve the throughput of data aggregation, data collection, and query algorithms, allowing the wireless sensor network to have a wider application space and better adaptability.

References

1. Michael Ekonde Sone (2015)," Efficient Key Management Scheme to Enhance Security-Throughput Trade-off *Performance in Wireless Networks", Science & Information Conference 2015* July 28-30.
2. Natasha Saini1 ,Nitin Pandey2, Ajeet Pal Singh(2015)," Enhancement Of Security Using Cryptographic Techniques", 978-1-4673-7231-2/15©2015 *IEEE.*
3. Takahiro Fujita, Kiminao Kogiso, Kenji Sawada, & Seiichi Shin (2015), "Security Enhancements of Networked Control Systems Using RSA Public-Key Cryptosystem", 978-1-4799-7862-5/15©2015 *IEEE.*
4. Yasmin Alkady, Mohmed I. Habib, Rawya Y. Rizk, (2013)" A New Security Protocol Using Hybrid Cryptography Algorithms", 978-1-4799-3370-9/13©2013 *IEEE.*
5. Bhushan Chaudhari, Prathmesh Gothankar, Abhishek Iyer, D. D. Ambawade (2012),"Wireless Network Security Using Dynamic Rule Generation of Firewall", *2012 International Confer-*

ence on *Communication, Information & Computing Technology (ICCICT)*, Oct. 19-20,2012.

6. Sangita A. Jaju, Santosh S. Chowhan (2015)," A Modified RSA Algorithm to Enhance Security for Digital Signature", 978- 1-4799-6908-1/15©2015 **IEEE.**

7. Ayman Tajeddine Ayman Kayssi Ali Chehab Imad El-hajj,(2014)" Authentication Schemes for Wireless Sensor Networks", *17th IEEE Mediterranean Electrotechnical Conference*, Beirut, Lebanon, 13-16 April 2014. 978-1-4799-2337-3/14©2014 IEEE.

8. Ashwak alabaichi, Adnan Ibrahem Salih (2015), "Enhance Security of Advance Encryption Standard Algorithm Based on Key-dependent SBox", ISBN: 978-1-4673-6832-2©*2015 IEEE.*

9. Kyung-Ah Shim,(2015)" A Survey of Public-Key Cryptographic Primitives in Wireless Sensor Networks" *IEEE Communications Survey & Tutorials*, Vol., No., 2012, 1553-877X (c) 2015 *IEEE.*

10. Madhumita Panda, Atul Nag (2015), "Plain Text Encryption Using AES, DES & SALSA20 by Java Based Bouncy Castle API on Windows & Linux", *2015 Second International Conference on Advances in Computing & Communication Engineering*, 978-1-4799-1734-1/15   ©   2015   IEEE   DOI 10.1109/ICACCE.2015.130.

11. S. Scott, R. Sylvia, K. Brad, et al (2003). Data-Centrie storage in sensor nets, *ACM SIGCOMM Computer, Communication Review*. 33(2003):137-142.

12. G. Abhishek, G. Jens, C. John (2003), Resilient data-centric storage in wireless ad-hoc sensor networks, *Proc. of the 4th Int'l Conf. on Mobile Data Management*, (2003):45-62.

13. Z. Wensheng, C. Guohong, L.P Tom (2003) , Data dissemination with ring-based index for wireless sensor networks, *IEEE Int'l Conf. on Network Protocols*, (2003):305-314.

14. G. Benjamin, E. Deborah, G. Ramesh, et al (2003). DIFS: a distributed index for features in sensor networks, Proc. Of the *1st IEEE Int'l Workshop on Sensor Network Protocols and Applications Anchorage.*(2003):163-173.

15. L. Xin, J.K Young, G. Ramesh, et al (2003). Multi-Dimensional range queries in sensor networks, *Proc. of the 1st Int'l Conf. on Embedded Networked Sensor Systems.*(2003):509-5l7.

16. R.H Wendi, C. Anantha, B. Hari (2000), Energy-Eficient communication protocol for wireless microsensor networks, *Proc.*

of the 33rd Hawaii Int'l Conf. on System Sciences. (2000):8020-8029.

Chapter 12

Challenges to Achieve Renewable Energy Targets

Ekta Mishra and Santosh Patani

Introduction

Energy has always been the key to man's greatest goals, and to his dreams of a better world. Energy is the basis of human life. The caveman started on the path of civilization when he discovered the energy in fire for light, and utilized the energy in his body to hunt for food and survival. Today, man has come a long, long way and discovered innumerable ways to make various forms of energy work for him. This quest for finding new uses of energy has led to exiting discoveries and inventions in fact, we cannot imagine a world without them. All activities, be it through machines or what we do or what is done to us is either a transfer of energy or the transformation of energy from one form to another. Machinery in our factories and farms, electricity for lighting and heating, petroleum for transportation, nuclear and solar power to aid exiting futuristic programs and inventions have all risen from man's interest in energy. **Today Electrical Energy has emerged as the basic need along with the Food, Clothing and Shelter.**

Technologically electrical energy is a vital infrastructure of economic development and so its demand is increasing day by day. Progress in the fields of industry, agriculture, communication, transport and other sectors is necessitating growing consumption of energy for developmental and economic activities.

Energy is one of the major inputs for the economic development of any country. In the case of the developing countries, the energy sector assumes a critical importance in view of the ever- increasing energy needs requiring huge investments to meet them. India in its Nationally Determined Contributions (NDCs)committed to three targets, which are to be achieved by the year 2030. First, by 2030, 40% of India's cumulative electric power installed capacity will come from non-fossil fuel-based energy sources. Second, India will reduce the emission intensity of its gross domestic product (GDP) by 33–35% (vis-à-vis 2005 levels). Third, India will create an additional carbon sink of 2.5–3 billion tonnes of CO_2 equivalent. India is the third-largest producer and second largest consumer of electricity in the world, with an installed power capacity of 382.73 GW as of March 2021. India was ranked fifth in wind power, fifth in solar power and fourth in renewable power installed capacity as of 2019. Figure1 shows the total installed capacity of India.

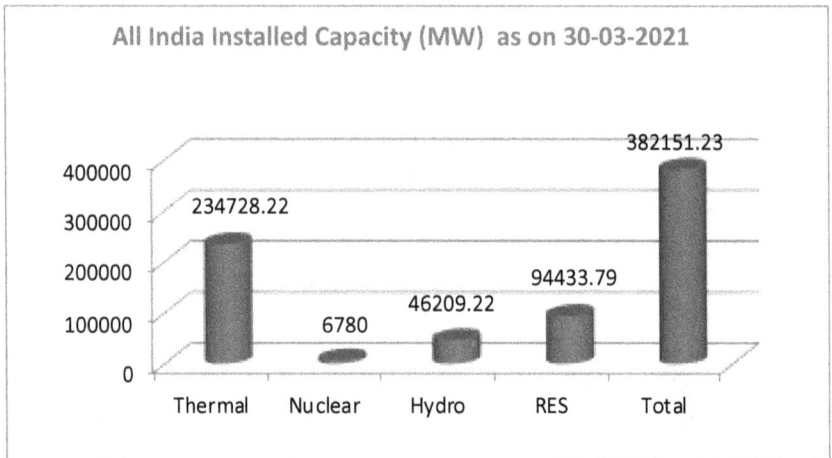

All India Installed Capacity (MW) as on 30-03-2021

Total installed capacity in India is contributed by following sources
1) Thermal
 (i) Coal
(ii) Gas and Lignite
(iii) Diesel
2) Hydro Power
3) Nuclear
4) Renewable Energy

Thermal	• Coal-India has large reserve of coal. The total installed capacity of coal by 30.03.2021 is 202.674GW • Gas Or Lignite-India gas and thermal power capacity stood 315.44 GW • Diesel-India diesel thermal power capacity is 0.509GW
Renewable	• Wind energy is the largest renewable energy source in India. Many renewable energy projects creates positive environment to wards investors to exploits India's potential. As of march 2021 the total installed capacity of renewable is 94.433 GW.
Hydro	• With large swathe of rivers and water bodies, India has enomorous potential of hydro power. The total installed capacity of hydro power as on march2021 is 46.209GW.
Nuclear	• As of march 2021 India had 6.78GW of total nuclear installed capacity . With one of the world's largest reserve of thorium India has huge potential in nuclear energy.

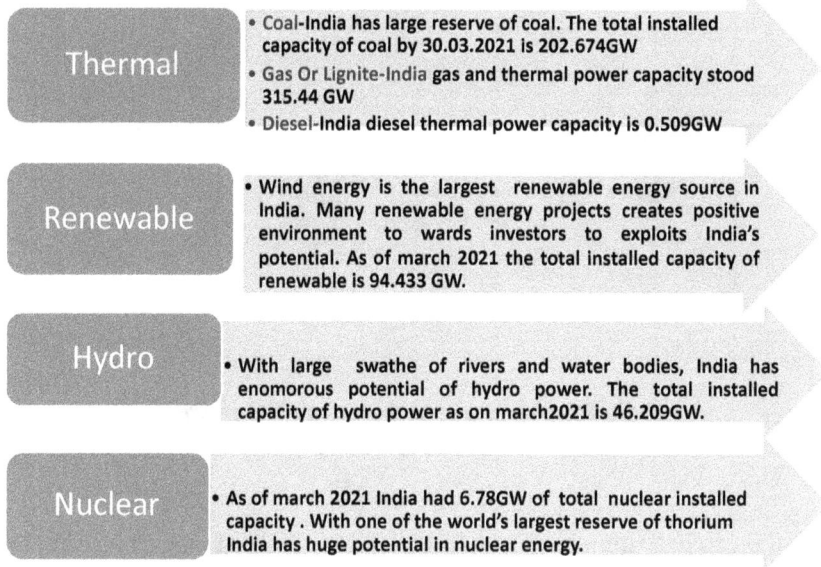

With a generation of 1,558.7TWh, India is third-largest producer and third-largest consumer of electricity in the world. Although power generation has grown more than 100 fold since independence, growth in demand has been even higher due to accelerating economic activity. India's energy firms have made significant progress in the global energy sector. Figure 2 shows the generation capacity of developed and developing country.

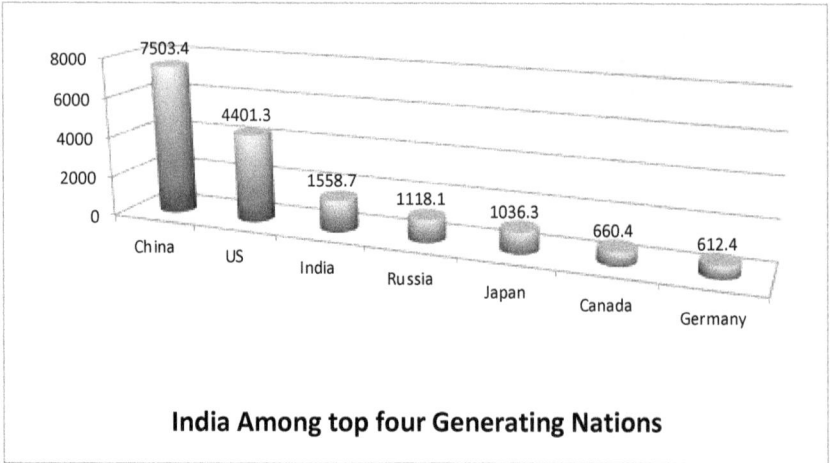

India Among top four Generating Nations

Total electricity generation of India for March 2021

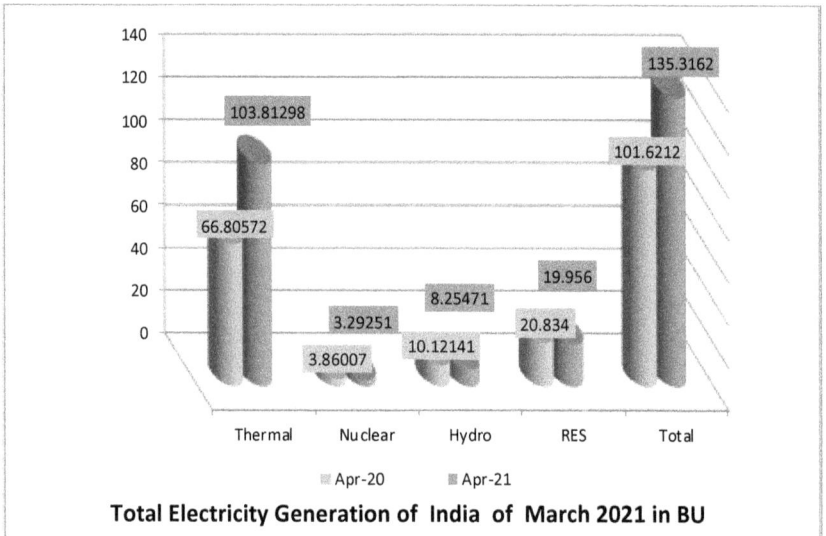

Total Electricity Generation of India of March 2021 in BU

188

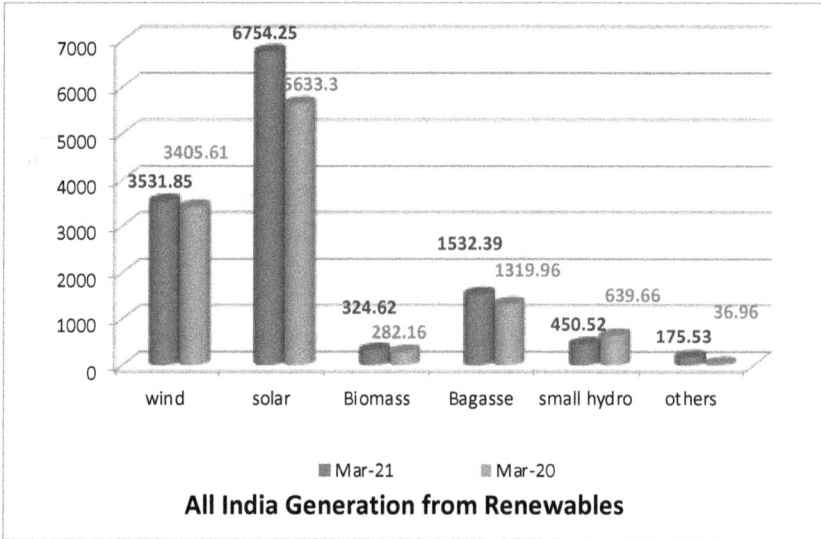

All India Generation from Renewables

Key benefits of Renewable Energy Sources

Time seems to be separated into a pre-COVID and a post-COVID period. Energy supply and demand have been disrupted, and carbon dioxide emissions fell. In such unprecedented times, stepping back to look at what happened in the renewable energy sector in 2019 may seem counter intuitive. But we need to do this. It's clear that we need to study the global picture with a long-term view to make the right decisions going forward. If we don't, we risk getting sidetracked by a short-term perspective. As disruptive as COVID-19 has been, the crisis does not alter observable trends in the energy sector that have persisted for years. The truth remains: we need to enact a structural shift built on an efficient and renewable-based energy system if we want to decarbonise our economies.

Like any human activity, all energy sources have an impact on our environment. Renewable energy is no exception to the rule, and each source has its own trade-offs. However, the advantages over the devastating impacts of fossil fuels are undeniable: from the reduction of water and land use, less air and water pollution, less wildlife and habitat loss, to no or lower greenhouse gas emissions.
In addition, their local and decentralised character as well as technology development generate important benefits for the economy and people.

- **Renewable energy emits no or low greenhouse gases. That's good for the climate.**

The combustion of fossil fuels for energy results in a significant amount of greenhouse gas emissions that contribute to global warming. Most sources of renewable energy result in little to no emissions, even when considering the full life cycle of the technologies.

- **Renewable energy emits no or low air pollutants. That's better for our health.**

Worldwide increases in fossil fuel-based road transport, industrial activity, and power generation (as well as the open burning of waste in many cities) contributes to elevated levels of air pollution. In many developing countries, the use of charcoal and fuelwood for heating and cooking also contributes to poor indoor air quality. Particles and other air pollutants from fossil fuels literally asphyxiate cities. According to studies by the World Health Organisation, their presence above urban skies is responsible for millions of premature deaths and costs billions.

- **Renewable energy comes with low costs. That's good for keeping energy prices at affordable levels.**

Geopolitical strife and upheavals often come with increasing energy prices and limited access to resources. Since renewable energy is produced locally, it is less affected by geopolitical crisis or price spikes or sudden disruptions in the supply chain.

- **Renewable energy creates jobs. That's good for the local community.**

The largest part of renewable energy investments is spent on materials and workmanship to build and maintain the facilities, rather than on costly energy imports. Renewable energy investments are usually spent within the continent, frequently in the same country, and often in the same town. This means the money citizens pay on their energy bill stays home to create jobs and fuel the local economy.

- **Renewable energy makes the energy system resilient. That's important to prevent power shortages.**

Renewables make urban energy infrastructures more independent from remote sources and grids. Businesses and industry invest in renewable energy to avoid disruptions, including resilience to weather-related impacts of climate change.

"An energy system based on distributed and decentralised generation is more flexible and resilient to those central shocks which are becoming more frequent with climate change"

- **Renewable energy is accessible to all. That's good for development.**

In many parts of the world, renewables represent the lowest-cost source of new power generation technology, and costs continue to decline. Especially for cities in the developing world, renewable energy is the only way to expand energy access to all inhabitants, particularly those living in urban slums and informal settlements and in suburban and peri-urban areas.

- **Renewable energy is secure. That's good for stability.**

Evolving energy markets and geopolitical uncertainty have moved energy security and energy infrastructure resilience to the forefront of many national energy strategies. Security of supply is a serious concern in energy markets worldwide, from the European Union and the United States to Egypt and India.

- **Renewable energy is democratic. That's good for acceptance.**

In recent years, the number of community energy projects using renewable sources has surged in various parts of the world. Although community energy is frequently associated with Northern European countries such as Denmark and Germany, such projects are emerging in other parts of the world including Thailand, Japan, and Canada. This trend confirms that democracy is an important driver for the change to renewable energy.

Major Challenges

There are many challenges to achieve renewable energy targets. some of them are

1) Higher upfront cost: The most obvious and widely publicized barrier to renewable energy is cost—specifically, capital costs, or the upfront expense of building and installing solar and wind farms. Like most renewables, solar and wind are exceedingly cheap to operate—their "fuel" is free, and maintenance is minimal—so the bulk of the expense comes from building the technology. Higher construction costs might make financial institutions more likely to perceive renewables as risky, lending money at higher rates and making it harder for utilities or developers to justify the investment.

2) Siting and transmission: Nuclear power, coal, and natural gas are all highly *centralized* sources of power, meaning they rely on relatively few high output power plants. Wind and solar, on the other hand, offer a *decentralized* model, in which smaller generating stations, spread across a large area, work together to provide power.

Siting is the need to locate things like wind turbines and solar farms on pieces of land. Doing so requires negotiations, contracts, permits, and community relations, all of which can increase costs and delay or kill projects.

Transmission refers to the power lines and infrastructure needed to move electricity from where it's generated to where it's consumed. Because wind and solar are relative newcomers, most of what exists today was built to serve large fossil fuel and nuclear power plants.

3) **Reliability Issues:**
Renewable energy technologies totally depend on the weather (e.g., sun and wind) to be able to harness any energy. In case atmospheric conditions are not good enough, renewable energy technologies would lack the ability to generate any electricity.

- Hydro generators require enough rain to fill dams for their supply of flowing water.
- Wind turbines require wind blowing, at least with minimum wind speed, to move their blades.
- Solar panels need clear skies and sunshine to get the heat required to generate electricity, and at night it isn't collected.

4) **Low-efficiency Levels**

Renewable energy technologies are still significantly new to the market, meaning, they still lack the much-needed efficiency. Lack of sufficient knowledge on how to effectively harness these forms of energy makes the installation and maintenance cost for such facilities quite high. This poses forecast problems, and investors may shy away from investing their money for fear of not getting returns pretty quickly.

5) **Expensive Storage Costs**
We often overlook the storage cost of renewable energy. In case of renewable energy, you must store the energy collected having a battery installed or else you will lose it.The overall storage cost for the energy is about 9 cents per kilowatt-hour; however, the cost of the battery is upfront.

6) **Difficulty in maintaining stable load flow graph**
Ideally, the daily load graph should be stable but because of uncertain nature of RE, at times it is very difficult to maintain the stability of load flow curve.

7) **Existing PPAs with the traditional power stations / generators**
As per the then scenario of power shortage, long term Power Purchase Agreements (PPA) were executed and for the survival of those power generators there is a clause of paying fixed cost even a single unit is not purchased.

8) **Market entry**
Most of the electricity demand is dominated by certain major players, including coal, nuclear, and, most recently, natural gas.Utilities across the country have invested heavily in these technologies, which are very mature and well understood, and which hold enormous market power.This situation—the well-established nature of existing technologies—presents a formidable barrier for renewable energy. Solar, wind, and other renewable resources need to compete with wealthier industries that benefit from existing infrastructure, expertise, and policy. It's a difficult market to enter. New energy technologies—startups—face even larger barriers. They compete with major market players like coal and gas, *and* with proven, low-cost solar and wind technologies. To prove their worth, they must demonstrate scale: most investors want large quantities of energy, ideally at times when wind and solar aren't available. That's difficult to accomplish, and a major reason why new technologies suffer high rates of failure.

9) **Policy and regulatory obstacles**

A comprehensive policy statement (regulatory framework) is not available in the renewable sector.the regulatory framework and procedures are different for every state because they define the respective RPOs (Renewable Purchase Obligations) and this creates a higher risk of investments in this sector.every state has different regulatory policy and framework definitions of an RPO. The RPO percentage specified in the regulatory framework for various renewable sources is not precise.

10) **Financial and fiscal obstacles**

The initial unit capital costs of renewable projects are very high compared to fossil fuels, and this leads to financing challenges and initial burden. There are uncertainties related to the assessment of resources, lack of technology awareness, and high-risk perceptions which lead to financial barriers for the developers.

11) **Technological obstacles**

MNRE issued the standardization of renewable energy projects policy which includes testing, standardization, and certification. They are still at an elementary level as compared to international practices. Quality assurance processes are still under starting conditions. The quality and reliability of manufactured components, imported equipment, and subsystems is essential, and hence quality infrastructure should be established. There is no clear document related to testing laboratories, referral institutes, review mechanism, inspection, and monitoring.

12) **Awareness, education, and training obstacles**

There is an unavailability of appropriately skilled human resources in the renewable energy sector. Furthermore, it faces an acute workforce shortage. After installation of renewable project/applications by the suppliers, there is no proper follow-up or assistance for the workers in the project to perform maintenance. Likewise, there are not enough trained and skilled persons for demonstrating, training, operation, and maintenance of the plant. There is inadequate knowledge in renewables, and no awareness programs are available to the general public. The lack of awareness about the technologies is a significant obstacle in acquiring vast land for constructing the renewable plant.the environmental benefits of renewable technologies are not clearly understood by the people and negative perceptions are making renewable technologies less prevalent among them.

13) **Environmental obstacles**

Large utility-scale solar plants require vast lands that increase the risk of land degradation and loss of habitat. The PV cell manufacturing process includes hazardous chemicals such as 1-1-1 Trichloroethene, HCL, H_2SO_4, N_2, NF, and acetone. A single wind turbine does not occupy much space, but many turbines are placed five to ten rotor diameters from each other, and this occupies more area, which include roads and transmission lines.

Conclusion:

It is observed that for human life, energy is one of the most important parameter. It has emerged as 4th basic need after food, shelter and clothing. We can not imagine modern life without energy. All the sectors like Agriculture, Industry, Education, Medical, Banking, etc are dependent on the energy.

The historical path of the development sector has seen many changes in last 110 years. The first step towards the legal framework was the 1910 Act, which authorised the licensees to generate electricity. Later on post-independence, the 1948 Act mandated the formation of State Electricity Boards which helped in bringing the electricity in the reach of common people. The turning point was the Electricity Act 2003. After this, there is noticeable growth in the energy sector. The Power shortage scenario up to the decade of 2010 has changed 180 degree and now by the end of 2020, we are now Power Surplus position.

Definitely, from the history it is eminent to notice that Development and Environment are always inversely proportional and if we opt for Development, Environment will be at risk and vice a versa if we opt for Environment, the Development will be slow.

In the energy sector, this is 100% true and recent changes in the climate have proved that CO_2 emission is a major threat for the climate. The studies and available data proves that, energy sector is major emitter of CO2, because most of the generation of electricity is from burning of fossil fuels.

Now, it is a major challenge to bring balance between Development and Environment. The possible solution is use of more and more Renewable Sources for electricity generation.

Accordingly, the global leaders and our own Government has designed policies to promote the Renewable Energy sources. In last few years the RE sector is emerging as new solution for creating win-win situation for all the stakeholders.

But still long way to go to achieve the desired results as there are some obstacles in the development of the RE sector in our country. The major obstacles are Technological, Financial &poor awareness. Definitely, we can achieve our goal if we initiate more and more research in bringing Technological advancement which will facilitate the manufacturing of equipment in house and at lower cost.

References

[1] Sanduleac, M.; Albu, M.; Stanescu, D.; Stanescu, C (2019). Grid Storage in LV Networks–An Appropriate Solution to Avoid Network Limitations in High RES Scenarios. In *Proceedings of the 2019 International Conference on Electromechanical and Energy Systems (SIELMEN)*, Craiova, Romania, 9–11 October 2019; pp. 1–6.
[2] Vai, V.; Alvarez-Herault, M.-C.; Raison, B.; Bun, L (2020). Optimal Low-voltage Distribution Topology with Integration of PV and Storage for Rural Electrification in Developing Countries: A Case Study of Cambodia. *J. Mod. Power Syst. Clean Energy 2020*, 8, 531–539.
[3] Cie´slik, S (2014). Voltage control in low-voltage distribution grids with micro-sources (in Polish: Regulacja napi͈ecia w sieciach dystrybucyjnych nn z mikroinstalacjami). In Proceedings of the Symposium, *Współczesne Urza͈dzenia Oraz Usługi Elektroenergetyczne, Telekomunikacyjne i Informatyczne'*, Poznan´ , Poland, 19–20 November 2014; pp. 24–27.
[4] Tang, J.; Cai, D.; Yuan, C.; Qiu, Y.; Deng, X.; Huang, Y (2019). Optimal configuration of battery energy storage systems using for rooftop residential photovoltaic to improve voltage profile of distributed network. *J. Eng.* 2019, 2019, 728–732.
[5] Zhang, Y.; Dong, Z.Y.; Luo, F.; Zheng, Y.; Meng, K.; Wong, K.P (2016). Optimal allocation of battery energy storage systems in distribution networks with high wind power penetration. *IET Renew. Power Gener.* 2016, 10, 1105–1113.
[6] H. Zhao, Q. Wu, S. Huang, Q. Guo, H. Sun, Y. Xue,(2015) "Optimal siting and sizing of Energy Storage System for power systems with

large-scale wind power integration", PowerTech, *2015 IEEE Eindhoven: IEEE, (*2015), pp. 1–6.

[7] Y.V. Makarov, P. Du, M.C. Kintner-Meyer, C. Jin, H.F. Illian, (2012) "Sizing energy storage to accommodate high penetration of variable energy resources", *IEEE Trans. Sustain. Energy 3* (2012) 34–40.

[8] P. Fortenbacher, M. Zellner, G. Andersson,(2016) "Optimal sizing and placement of distributed storage in low voltage networks", Power Systems Computation Conference (PSCC), 2016: *IEEE*, (2016), pp. 1–7.

[9] S. Bahramirad, W. Reder, A. Khodaei,(2012) "Reliability-constrained optimal sizing of energy storage system in a microgrid", *IEEE Trans. Smart Grid 3* (2012) 2056–2062.

[10] I. Miranda, N. Silva, H. Leite (2016), "A Holistic Approach to the Integration of Battery Energy Storage Systems in Island Electric Grids With High Wind Penetration", *IEEE Trans. Sustain. Energy 7* (2016) 775–785.

[11] P. Hu, R. Karki, and R. Billinton(2009), "Reliability evaluation of generating systems containing wind power and energy storage," *IET Generation, Transmission & Distribution,* vol. 3, no. 8, pp. 783-791, Aug. 2009.

[12] F. A. Bhuiyan and A. Yazdani,(2010) "Reliability assessment of a wind-power system with integrated energy storage," *IET Renewable Power Generation,* vol. 4, no. 3, pp. 211-220, May 2010.

[13] Z. Yi, Z. Songzhe, and A. A. Chowdhury (2011), "Reliability Modeling and Control Schemes of Composite Energy Storage and Wind Generation System With Adequate Transmission Upgrades," *IEEE Transactions on Sustainable Energy,* vol. 2, no. 4, pp. 520-526, Oct. 2011.

[14] S. A. Arefifar, Y. A. I. Mohamed, and T. H. M. El-Fouly (2012) "Supply- Adequacy-Based Optimal Construction of Microgrids in Smart Distribution Systems," *IEEE Transactions on Smart Grid,* vol. 3, no. 3, pp. 1491-1502, Sept. 2012.

[15] S. A. Arefifar, Y. A. R. I. Mohamed, and T. H. M. El-Fouly, (2013) "Comprehensive Operational Planning Framework for Self-Healing Control Actions in Smart Distribution Grids," *IEEE Transactions on Power Systems,* vol. 28, no. 4, pp. 4192-4200, Nov. 2013.

Chapter 13

DESIGN AND SIMULATION OF TEXTILE ANTENNA FOR WEARABLE APPLICATIONS

M. Pandimadevi[1], R. Tamilselvi[2] and M. Parisa Beham[2]
Department of ECE, Sethu Institute of Technology, India[1]
Department of ECE, Sethu Institute of Technology, India[2]

ABSTRACT
Usage of Flexible antennas for the advancement of wearable gadgets have been broadly expanded because of less weight, superior, minimized structure and simple manufacture. The proposed antenna is designed with rectangular patch with slot using a Textile substrate material such as Jute. The various performance parameters such as return loss, Voltage Standing Wave Ratio (VSWR), directivity etc., are obtained. The frequency range is chosen as Industrial Scientific and medical application band (ISM) frequency range of 2.46 GHz. The results show that patch antenna with jute substrate has better results in terms of return loss, VSWR and Directivity both in flat and on-body conditions. Results demonstrate the suitability of this flexible antenna for on-body wearable communications.

1. INTRODUCTION

Body-centric communications takes its place firmly within the sphere of personal area networks (PANs) and body area networks (BANs). One of the applications – the on-body communications – It is the link between bodies mounted devices communicating wirelessly, while off-body communication is the radio link between body worn devices and base units or mobile devices located in surrounding environment. Finally, in-body communication is communication between wireless medical implants and on body nodes [1]. Patch antenna has many advantages, like low weight, low profile, easy to fabricate, integrated into microwave integrated circuits [2]. The limitations of conventional microstrip patch antennas are more deposition of electromagnetic signals in the human body that is high specific absorption rate (SAR) though their physical size is large. Secondly, due to size it is difficult to integrate and make them hidden inside the clothing of wearer [3-9]. Hence, a wearable antenna that is the textile based antenna is one of the better alternatives for such type of applications. The wearable antenna should be light weight, flexible, compact, and hidden and should be easily integrated within the clothing and it should not affect the health of wearer.

There are several antenna types in our modern world. For example, Yagi Uda, Loop, parabolic and Microstrip Patch antenna. Each of them has their own advantages and disadvantages. Among them, Microstrip patch antenna is recently used in the design of flexible antenna. With increasing requirement for body area network, personal area network and mobile communications, the need for smaller and low profile antenna has brought the MPA to the forefront. Some of the principal advantages of microstrip antennas compared to conventional microwave antennas are [10]:

- Light weight and thin profile configurations, which can be made conformal to shaped surfaces;
- Cheaper;
- Easily integrated with microwave integrated circuits;
- Easy to fabricate.

The rest of the paper has been organized as follows: Section 2 explains survey of various existing methodologies available in the flexible and wearable antenna. Section 3 deals with the design of the proposed antenna. Section 4 deals with the bending results and discussions. Section 5 deals with the conclusion and future work.

2. LITERATURE SURVEY

Prior to the antenna design, it is important to know the existing methodologies and its issues.

Sonia C. Survase [11] et al designed micro strip yagi patch antenna for telemedicine application. Two types of yagi uda antenna: circular and rectangular were designed having gain 2dB and 7.8dB respectively in HFSS software. The operating frequency is 2.54GHz. In this paper, reasonable return loss (-24dB) and VSWR (1.1) was achieved, but there is no evidence of on-body applications of the antenna.

K.S.Chakradhar et al[12] designed a rectangular microstrip patch antenna and then rectangular with U-slot antenna with slits.. An insulating denim fabric with dielectric constant 1.7 with a thickness of 0.7 is used for preparing the substrates. The designed antenna is simulated using Ansoft HFSS Software. The results shows that the rectangular patch resonate at 2.5 GHz with the return loss of -16.86dB and the U-slot Antenna resonated at 2.2 GHz with return loss of -41.68dB and 3.9 GHz return loss of -16. 16dB.But the dimension of the antenna is very large as 120mmx120mm.

Saadat Hanif Dar et al [13] proposed patch antenna on a rubber substrate and its performance near human body is investigated. Simulations were done using CST studio suite. It is observed that bandwidth and return loss are improved by adding rubber contents. The antenna efficiency and its gain was decreased, but this was considered reasonable since natural rubber is quite loss in nature.

Liu Jianying, Dai Fang et al [14] proposed a flexible Yagi-Uda antenna with polyimide substrate for wireless body area network (WBAN). The prototype covers 2.45GHz Industrial, Scientific and Medical (ISM) band. The effects of bending along x-axis and y-axis for flexible Yagi-Uda antenna is analyzed. The resonant frequency, bandwidth and radiation patterns are simulated with various bending conditions. Results indicate that with increasing the bending degree, the input matching become worse. It is worth noting that the maximum radiation orientation at H-plane of endfire Yagi-Uda antenna deviates with bending along y-axis.

Shenbaga K et al [15] designed a wearable antenna with PDMS+glass as substrate material for medical applications. The operating frequency is 2.8GHz. Return loss obtained is -19dB.But the material is very expensive when compared with commercially available substrate materials. The antenna is also very rigid, very hard to bend.

3. ANTENNA DESIGN

The proposed micro strip antenna was designed with ground plane of copper sheet of thickness 0.035mm thickness having dimensions 106x106 mm as shown in figure 1. Above the ground plane was the substrate layer of same dimension as that was of ground plane having thickness 1 mm. Jute is used as substrate material. The square patch of 53mm side is chosen with 2 slots of 42mm length and 10 mm wide. The strip line of 25mm length with 9 mm wide was excited from port 1 via a 50Ω coaxial probe transmission line.

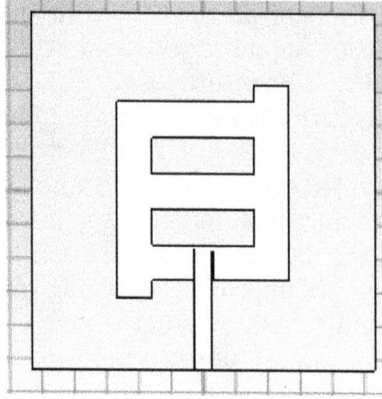

Fig.1 Proposed antenna design

4. ANTENNA SIMULATION

The design was simulated in Computer Simulation Technology (CST) Studio suite software. Figure 2 shows the proposed design using Jute substrate in CST software. The Return loss, VSWR, Radiation pattern and directivity measurements are shown in figures 3,4, 5 and 6 respectively.

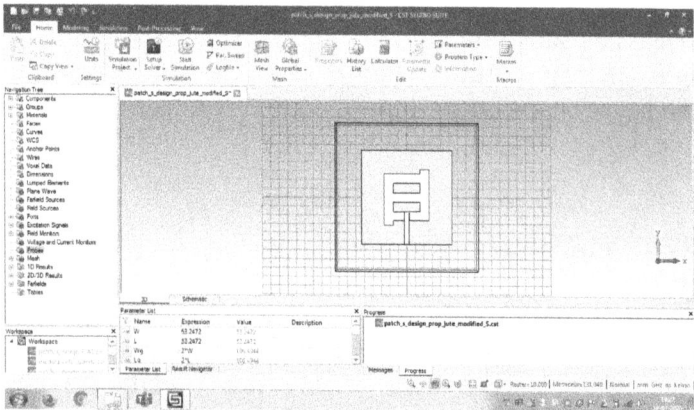

Fig.2 Design of antenna using Jute substrate

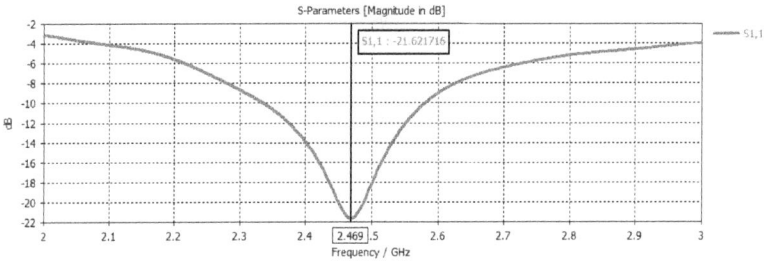

Fig.3 Return loss measurement

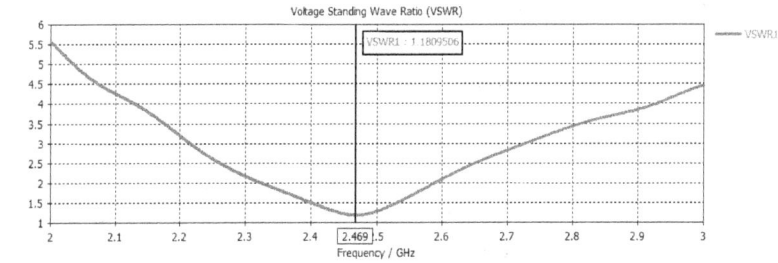

Fig.4 VSWR measurement

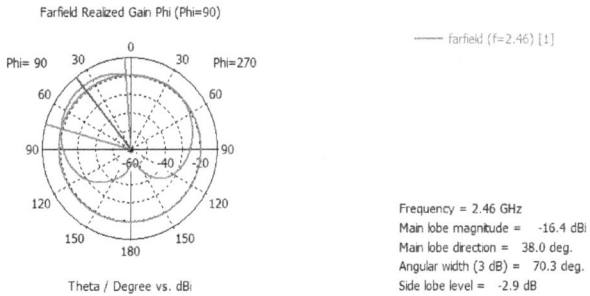

Frequency = 2.46 GHz
Main lobe magnitude =   -16.4 dBi
Main lobe direction =   38.0 deg.
Angular width (3 dB) =   70.3 deg.
Side lobe level =   -2.9 dB

Fig.5 Radiation Pattern

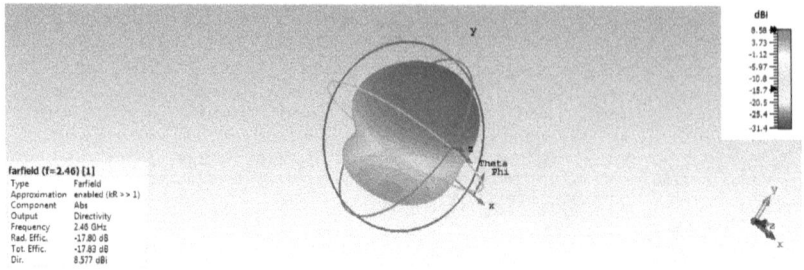

Fig. 6 Directivity measurement

5. RESULTS AND DISCUSSION

The proposed antenna using jute substrate material was designed and simulated. Antenna have shown better results and the values of various parameters are tabulated in Table 1. The return loss is achieved with -21.62dB. The Voltage Standing Wave Ratio (VSWR) is achieved with 1.18 and directivity also achieved a high value of 8.57dBi. These results show that the jute substrate is well adaptive with better return loss, VSWR and directivity at Industrial Scientific and Medical band frequency (2.45-2.55GHz). Since the jute fibre is a biodegradable, cheap, flexible, nontoxic, environment-friendly, strongest fibre [16][17], compatible, higher thermal stability and easily dryable after wash and recyclability, it is the correct choice as a substrate for the fabrication of microwave antennas. Since the substrate material is flexible, it can be bent

to any degree and the substrate is a textile material, the antenna can be stitched to any cloths and the results made the antenna suitable for wearable and ISM band applications.

Table I. Various parameters of proposed antenna

Substrate Material	Operating frequency(GHz)	Return loss(dB)	VSWR	Directivity (dBi)
Jute	2.469	-21.62	1.18	8.57

204

6. CONCLUSION AND FUTURE WORK

The proposed antenna presented are very versatile and it is easy to make it operate at various frequency bands. In addition, antenna using jute substrate will improve the bandwidth thus suitable for wearable applications such as military, medical, industrial etc., The results were also satisfactory at 2.46GHz. In future, the proposed antenna will be fabricated and measured under all conditions and will make them suitable for commercial applications.

References

[1] Hall, P. S., and Hao, Y(2006)., "Antennas and Propagation for Body Centric Communications", *European Conference on Antennas and Propagation (EuCAP)*, November 2006

[2] Baker-Jarvis ; Janezic, M.D.; DeGroot, D.C (2010). High-Frequency Dielectric Measurements. *IEEE Trans. Instrum.* Meas. 2010, 13, 24–31.

[3] Sonia C. Survase,Vidya V.Deshmukh (2013) "Design of Wearable Antenna for Telemedicine Applicatio",*IJESIT*, Volume 2, Issue 2, March2013.

[4] Rita Salvado 1,*, Caroline Loss 1, Ricardo Gonçalves 2 and Pedro Pinho (2012) Textile Materials for the Design of Wearable Antennas: A Survey ",*Sensors2012*, 12, 15841-15857; doi:10.3390/s121115841.

[5] S.Sankaralingam and Bhaskar Gupta (2009) "A Circular Disk Microstrip WLAN Antenna for Wearable Applications" 2009 *Annual IEEE India Confere*nce, 18-20 Dec. 2009 DOI: 10.1109/INDCON.2009.5409355.

[6] Balanis, C.A. Antenna Theory: Analysis and Design (2005), 3rd ed.; *Wiley Interscience*: Hoboken, NJ, USA, 2005.

[7] Morton, W.E.; Hearle (2008), W.S. Physical Properties of Textile Fibres, 4th ed.; *Woodhead Publishing: Cambridge*, UK, 2008.

[8] Locher, I.; Klemm, M.; Kirstein, T.; Tröster, G (2006). Design and Characterization of Purely Textile Patch Antennas. *IEEE Trans. Adv. Pack.* 2006, 29, 777–788.

[9] Abid Ahmad, Muhammad Farooq, Gulzar Ahmad,Muhammad Amir,Arbab Masood (2018),". A Review on Properties Amelioration of Wearable Antennas ", *International Journal of Engineering works*, Vol. 5, Issue 5, PP. 111-115, May 2018.

[10] Indrasen Singh, Dr. V.S. Tripathi, (2011)" Micro strip Patch Antenna and its Applications: *a Survey",IJCTA*, Sept-Oct 2011.

[11] Sonia C. Survase, VidyaV.Deshmukh (2013)," Design of Wearable Antenna for Telemedicine Application*", International Journal of Engineering Science and Innovative Technology (IJESIT)* Volume 2, Issue 2, March 2013 .pp.574-580.

[12] K.S.Chakradhar, Inumula Veera raghava Rao,DT.Durga Prasad,P.Raju, V. Malleswara Rao (2019)," Wearable Textile Patch Antenna for Medical Applications", *International Journal of Innovative Technology and Exploring Engineering* IJITEE)ISSN: 2278-3075, Volume-9 Issue-2S3, December 2019.

[13] SaadatHanif Dar, Jameel Ahmed and Muhammad Raees, (2016)" *Characterizations of Flexible Wearable Antenna based on Rubber Substrate*", (IJACSA) *International Journal of Advanced Computer Science and Applications*, Vol. 7, No. 11, 2016.

[14] Liu Jianying, Dai Fang*, Zhang Yichen, Yu Xin, Cai Lulu, Zuo-Panpan, and Wang Mengjun (2016),"*BendingEffects on a Flexible Yagi-Uda Antenna for Wireless Body Area Network*", 7th Asia *Pacific International Symposium on Electromagnetic Compatibility* 2016.

[15] Shenbaga K, Takshala Devapriya A (2019)," Wearable Antenna for Medical Application", *International Research Journal of Engineering and Technology(IRJET)*, Volume: 06 Issue: 04 ,pp-2175-2179, Apr2019.

[16] Shahinur, S., Hasan, M., Ahsan, Q., Saha, D. K., & Islam, M. S. (2015),"Characterization on the Properties of Jute Fiber at Different Portions", *International Journal of Polymer Science,* pp:1–6,2015, doi:10.1155/ 2015/262348.

[17] Basu, G., & Roy, A. N (2018).,"Blending of Jute with Different Natural Fibres. *Journal of Natural Fibers*, 4(4), pp:13–29,2018, doi:10.1080/15440470801893323.

Chapter 14

Research Issues and Challenges of Embedded Systems

R. Banupriya
Department of Electrical and Electronics Engineering,
PGP College of Engineering &Technology, Tamilnadu, India

J. Manjushree Kumari and C Jisha Chandra
Department of Electrical and Electronics Engineering,
Gnanamani College of Technology, Tamilnadu, India

Abstract:

The emphasis of embedded system engineering is often on resource efficiency, cost reduction, and functionality. If non-functional qualities are taken into account during the development of embedded systems, they are primarily concerned with safety and performance. Due to cost or efficiency constraints, security problems are ignored, incorporated as an afterthought, or downplayed. One reason for this is that many embedded devices formerly only had extremely limited connection and only worked in controlled situations. As a result, physical shielding of embedded devices was typically adequate for security. The situation has shifted radically. The evolution of embedded systems to devices linked through the Internet, wireless communication, or other interfaces, as well as the tendency toward ever-increasing numbers of devices (Internet of Things), necessitates a rethinking of embedded systems engineering procedures. Adding security measures late in the development process is no longer viable to reach the needed degree of security. Security engineering must be a part of the development process at all levels.

1. Introduction

Embedded devices, in terms of security, are systems that belong to a certain entity and are used in a potentially hostile environment. As a result, criteria that could be believed to be addressed by systems using physically secured devices must be reconsidered. Non-repudiation of data, time or status of devices managing sensor networks, regulatory requirements for calibration and gauging devices, payment system security, or the enforcement of a certain behaviour of remote-controlled equipment are all possible instances. For such devices, a number of international and national rules and standards apply. Many of them need anti-manipulation protection. Many technological fields, including automotive, avionics, telecommunications, factory automation, medical systems, and consumer electronics, rely on embedded systems. Embedded devices provide money for their owners and play crucial roles in critical situations, hence their proper operation is critical. Communication lines, as well as embedded devices themselves, are becoming more vulnerable to assaults that have previously been seen in other IT systems. Traditional security measures such as physical isolation from harmful situations are no longer enough. One reason is that an increasing number of embedded devices have communication interfaces and communicate over open network infrastructures (e.g., the Internet, wireless, peer-to-peer), and helpful advances in user control, such as convenient browser-based configuration, introduce new attack vectors, especially if the devices are used in systems of systems. Reliable cooperative reasoning in networked embedded devices with connections to information infrastructures may lead to new security needs in systems of systems.

Mobility opens up new attack possibilities, such as malware being physically transported from one portion of a system to another, bypassing firewalls and other security measures. Thus, device identification, control over a device's operational state, security of device communication, and even non-repudiation of actions taken at a specific time are security requirements that are becoming increasingly important in today's embedded scenarios; at the same time, existing hardware and software combinations do not support these requirements. Furthermore, with the exception of mobile platforms, security considerations are often overlooked in contemporary embedded system research and development efforts. Plant automation, remote metering, and essential communication situations all rely on the secure and dependable functioning of each individual embedded component. Furthermore, Future Internet thinking entails thinking beyond certain pre-planned intelligent device applications.

An electronic or computer system was created with the intention of controlling and accessing data from electronic-based systems. A single chip microcontroller (cortex, ARM, etc.) plus a microprocessor make up a general embedded system (FPGAs, DSPs, and ASICs). Because of their versatility and flexibility, these technologies formed the core of a vehicle's electronic system. Microcontrollers, DSPs, or both are widely referred to as Electronic Control Units in automobiles all around the world. Nowadays, a large number of embedded controllers are used in a variety of standard and premium automobiles. A microprocessor that controls an engine is an example of an embedded system, which is a decisive combination of hardware and software that becomes a vital component for higher machines. Nonetheless, hardware constraints like as memory, battery charge, and processor power dominated this system, resulting in low to moderate programme complexity. An embedded system is one that is meant to function without human intervention and may be required to be responsive to real-time occurrences. High speed, low power dissipation, compact weight and size, accuracy, and long-term reliability are key requirements for embedded systems. The fastest growing sector in embedded applications is networked embedded systems.

2. Embedded Systems

Embedded systems have had a significant impact on the automobile industry since its inception. Fuel injection and combustion controller devices, airbags, event data recorders, anti-lock braking system, adaptive cruise control, black box, drive by wire, satellite radio, telematics, traction control, automatic parking, entertainment systems, night vision, heads-up display, back up collision sensors, navigational systems, tyre pressure monitor, climate control, and other embedded systems are among the most commonly used in vehicles. The advanced application of embedded systems in the vehicle is for pollution control and system monitoring. M2M or V2V communication, which is the critical assistance from a temporary network, and traffic management and forecasting systems established in metro cities, seamlessly gather information from many sources to assist drivers and traffic administration. Only a portion of the vehicle and network, namely the embedded computer and communication systems, can provide real-time management. Customer satisfaction has increased as a result of the widespread adoption of vehicle and fleet monitoring, while opex and downtime have decreased. Furthermore, for multimedia and infotainment networking, a media oriented systems transport (MOST) system provides an efficient and cost-effective means of transmitting and controlling data between devices, even in the harsh environment of a car. Embedded technologies are already being used by several car manufacturers to create autonomous vehicle control. Electromobility and vehicle connection to smart phones and infrastructure will become a reality in a critical development where advanced driver assistance systems (ADAS) and autonomous automobiles have evolved. The advancements in vehicle electronic architecture have posed issues for the electronic design automation (EDA) and embedded systems communities in terms of vehicle embedded system design, security, and authentication.

India, as a growing nation, needs to upgrade its existing transportation systems and road infrastructure in order to improve current and future traffic flows, mobility, and safety. Intelligent Transportation Systems (ITS) are a cutting-edge application that may improve productivity, safety, and environmental performance when used on transportation and infrastructure to exchange data across frameworks. Parking Guidance and Information (PGI) systems and Parking Reservations (PRS) systems are the most sophisticated ITS applications, since they seamlessly connect live information and input from various sources. PGI systems have been shown to be more helpful than traditional parking since they provide vehicles with quick information regarding parking availability, cost, and navigation. The potential for PGI to increase motorist competition for a suitable parking place may be mitigated by reservations (PRS systems) or guaranteed parking spaces. This is because PRS systems take into account the goals of both drivers and parking management. As a result, good parking resource usage will improve parking income while reducing traffic congestion. Furthermore, cameras attached to a leading car detect traffic light colour variations, which are sent on to following vehicles, allowing them to adjust their speed to prevent collisions. Such camera observations may be used to identify meteorological conditions. By creating real-time meteorological micro-maps, this allows adjacent cars to avoid low-visibility zones. By comparing the local reference database, the identification of licence plates and their projected GPS coordinates may be sent to police agencies for automated location. Smart bike tracking systems have been created to assist prevent bike theft and save lives. This uses a car monitoring system to identify an accident and sends an SMS alert to the surrounding area. By simply forewarning the driver of its speed restrictions and crucial area detection, embedded systems have been successful in avoiding the challenge of managing vehicle real-time speed. As a consequence of an implanted intelligence inside itself, a car may propel itself without human input. Due to the need for increased security, the differential Global Positioning System (GPS) has proven to be a versatile and reliable system that quickly handles with selective availability and satellite clock problems. In the realm of electric cars, vehicle battery technology has a strong reputation. In terms of embedded systems design and software, their use and administration, in particular, are posing new challenges. The rate-capacity effect affects the power capacity of batteries throughout the charging process, making cruising range one of the most important criteria for electric cars. Furthermore, embedded systems provide crucial activities such as power flow regulation and battery management in the charging infrastructure to transport electrical energy from the grid to the EV.

The rapid development of embedded systems makes the transition from manual to automated systems much easier. Embedded has become one of the most important parts of any contemporary vehicle's electrical architecture because to its extreme flexibility and adaptability. Since the previous decade, embedded advancements have affected the key characteristics of car design. Following that, it will be required to focus more on the domains of electric and autonomous cars in terms of their real-world surroundings, security, and energy efficiency, since these are the most exciting sectors for embedded systems in the automotive industry. The software coded inside the microcontroller can also address a limited range of issues, which has been identified as an important area for embedded research. The complexity of very scattered and varied E/E architectures in today's automobile industry has resulted in an increased focus on ECU alliance. Despite the fact that established embedded system technologies have progressed significantly, several emergent issues of dependability and speed from the industry have created opportunities for new researchers.

An embedded system is made up of a mix of electrical components, computer hardware, and software. Between computer hardware and software, the electronic system serves as a bridge. The electrical system acts as a controller between the computer programme and the computer hardware. The complete assembly might be programmable or non-programmable, depending on the applications it will be used for. The embedded system provides the foundation for the current catchphrase Internet of Things, which has inspired people all over the globe to automate practically every aspect of their lives. Through its amazing uses, this breakthrough has catapulted the globe into a massive revolution. The fundamental concept of an embedded system is a method of functioning, organising, and completing a single or numerous tasks in accordance with a set of rules. The set of rules is just a software created for a certain system that causes all of the components to come together and function together. The applications of embedded systems are many, ranging from home domestic peripherals to critical defensive weapons. Medical, genetics, manufacturing, entertainment, visualisation, space research, bio informatics, and other sectors have recently begun to integrate embedded systems to improve their daily operations. These embedded systems are either microcontroller or microprocessor based, allowing us to reach complete functionality.

Embedded systems have several hidden benefits that make them more appealing in almost every sector. Massive amounts of data are depleting knowledge, and automated data gathering techniques are proving to be incredibly useful and efficient.

Once the system is in place, it may capture and transmit a large quantity of essential data for later use in data analytics and analysis. Embedded systems are becoming increasingly popular and useful because to their speed, size, power, durability, precision, and adaptability. They are suited for high-precision tasks in the healthcare and defence domains because to their speed and accuracy. Because of their small size, they're ideal for supporting smartness in portable and mobile devices. This clever system's lower power usage is an extra benefit.

3. Embedded Systems Applications

We are now entering an era of totally autonomous environments, when everything will be completed in a fraction of a second and without the need for human intervention. The increased usage of embedded systems in all disciplines has enabled this. Embedded computers are becoming pervasive throughout our life, from automobiles to cell phones, video equipment to mp3 players, and dishwashers to house thermostats. These systems are the engine that propels technological progress in every aspect of our lives. Embedded systems began with iPods, MP3 players, Bluetooth headsets, and PlayStations, and have since expanded to include washing machines, smart phones, self-driving vehicles, banking, military, space, research, and defence. Concerns concerning embedded systems' security and privacy have increased drastically in a short period of time as their usage has grown.

Embedded systems are the driving force behind technical advancement in a variety of fields, including automation, industrial monitoring, and control systems. As more computational and networked devices become widespread and "invisible" in our lives, security becomes more important for the reliability of any smart or intelligent systems based on these embedded systems. Assuming that embedded devices are not susceptible to cyberattacks, that embedded devices are not appealing targets for hackers, and that embedded devices get appropriate security via encryption and authentication is a significant error for any organisation. This article discusses some of the security concerns and threats that are connected with embedded systems, as well as the problems that they face and some of the solutions that have been offered.

Embedded system software is fixed and has limited flexibility, allowing users to write and execute their own programmes. It's a computer system designed for a particular use that's integrated into a broader mechanical or electrical system. It does a certain job again and over again in accordance with the specified instructions. It is made up of a mix of software and hardware, as well as mechanical pieces if necessary. It's often due to a real-time computer system limitation. As a result, an embedded system is a system that is controlled by a computer that is housed inside the system. Embedded systems are employed in a variety of applications depending on the needs, but their structure and operating principle in terms of system hardware and design process are the same. Extra hardware implementation, such as standard input and output devices, may be required for certain applications, such as mechanical and chemical plants, although it is not required for all plants and other devices. Embedded systems are often built around a microcontroller, with memory, timers, input/output ports, and counters all incorporated into the CPU, eliminating the need for additional memory. These may be divided into three groups based on their size: tiny, medium, and giant. Embedded systems aren't stand-alone devices; they're part of a larger gadget.

A network of embedded computers installed in the real world that interacts with the environment is known as an embedded sensor network. These embedded computers, also known as sensor nodes, are typically tiny, low-cost computers with a variety of sensors and actuators. These sensor nodes are deployed in situ, meaning they are physically put near the items they are perceiving. An embedded system is a sort of computer system that executes a set of pre-defined programmes and is often used in conjunction with a larger electrical or mechanical system. It usually starts with small MP3 players and progresses to highly complex hybrid vehicle systems. Keyboard, mouse, ATM, TV, PDA, cell phone, printer, elevator, smoke detector, DVD player, refrigerator, camera, GPS navigator, radio, TV remote, telephone, game controller, monitor, digital image processor, bar code reader, SD card, washing machine, anti-lock breaking system, blender, and so on are some examples of commonly used embedded systems in our daily lives. We employ embedded systems because of their durability, efficiency, and ability to satisfy real-time requirements. In terms of utilisation, examples of embedded systems indicate that they have become an integral part of our everyday lives. Because we have a smart embedded system in our house, we are extremely acquainted with the phrase "Smart Home." Almost all embedded systems nowadays are linked to the internet. Since a result, security risks have become a significant concern in recent years, as most embedded systems are even less secure than personal computers. One of the causes for this lack of security is that embedded system vendors have extremely limited hardware and software implementation choices. They must again contend with the competitive market price of other embedded manufacturer companies, as they all strive to maintain the lowest possible price in order to maintain customer satisfaction, while also failing to conduct any specific security research on their manufactured embedded products. This is a security risk for embedded devices, since maintaining advanced security measures for embedded systems entails a greater price for such embedded items. Customers, in general, do not want to pay extra for an embedded device, and they are unconcerned about potential security dangers to their goods. Lack of security analysis and manufacturing businesses' low-cost market product mentalities provide the hackers with the precise atmosphere they want. Many hacking tools for embedded systems may be found on the internet. PDA and modem hacking are two popular examples of embedded system hacking. Because of its applications in the TCP/IP protocol for inter-media interfacing, recent embedded systems protocol development trends are going to be convergence. Using IPv6 for embedded application development will be substantially more expensive in this instance, at least for the next several years. As a consequence, IPv4 will predominate in embedded system applications. Internal security difficulties in terms of authentication, integrity, and secrecy make IPv4 considerably more difficult.

Embedded device topologies are often the same in terms of system components and design approaches, despite the fact that there are many different kinds of applications. Standard I/O (Input/Output) devices may be required for complex applications such as chemical plants, although they are not required for most other embedded systems. Microcontrollers are used in the majority of embedded systems nowadays, which means memory and other particular devices are integrated into the Central Processing Unit (CPU). It is often classified into three sizes: small, medium, and giant. 4-bit microcontrollers are required for small applications such as TV remote controls. Microcontrollers with 8-bit or 16-bit are sufficient for medium-sized systems such as automated data collection systems, while 32-bit or more are required for high-end large-scale computer systems such as plant monitoring and central control. Embedded systems are seldom utilised on their own; instead, they are often employed as part of a larger, more complicated device. For the usefulness and safety of such devices, performance-based real-time constraints must be satisfied. For tiny size devices such as basic buttons or LEDs, a graphical user interface is not usually required (Light Emitting Diode). However, it is required for larger and more complicated devices, such as nuclear power plant systems, networks, data bus connections, and screen-edge systems, among others.

4. Conclusion

Embedded systems, in general, are intended to do any pre-defined activity that must be completed within a certain amount of time. The major distinction between a computer and an embedded system is that a computer may do many jobs that are specified by the user. An embedded system, on the other hand, is utilised to fulfil a certain duty that has been pre-defined by the producers. An embedded system's ability to satisfy all of the real-time requirements is a critical feature. There are two aspects to a real-time restriction. There are two types of real-time systems: hard real-time and soft real-time. In a hard real-time system, all deadlines must be met with zero flexibility, but in a soft real-time system, some flexibility is tolerated. For embedded devices, it is not necessarily required to be freestanding. The majority of embedded systems are really part of a larger computerised equipment. MP3 players, cameras, and TV remote controls are examples of freestanding embedded devices. A automobile and a nuclear power plant are two excellent examples of integrated embedded devices. GPS, fuel injection controller, anti-lock braking system, transmission controller, cruise control, active suspension, air-bag system, air-conditioner, and display monitor are all part of a contemporary automotive system. The software instructions developed for embedded systems are referred to as "firmware." It's kept in read-only memory (ROM) or a flash memory chip. Computer gear, for example, does not need a lot of resources to operate. The specialised user interface is another significant feature of embedded systems. It might be anything from a simple text-based user interface to a complicated graphical user interface. There is no requirement for a user interface with a basic button and LED system. The duty of a button may vary with the on-screen display, and the user has control over the option. A excellent example of a user interface system is a handheld device such as a joystick that must be aimed towards the screen. For an embedded device, the size and weight should be reduced. Microcontrollers are thus employed in embedded devices to provide the highest performance on demand. Microcontrollers are often needed to execute repetitive tasks over extended periods of time without failure. Aside from that, it must be dependable and safe in the case of certain unique systems, such as anti-lock braking systems in automobiles and nuclear power plant control systems. Embedded systems must also be cost effective in addition to these attributes. Manufacturers strive to keep their product prices as low as possible. It may also be linked to the physical world via sensors and actuators.

References

1. Michael Ekonde Sone (2015)," Efficient Key Management Scheme to Enhance Security-Throughput Trade-off Performance in Wireless Networks", *Science & Information Conference* 2015 July 28-30.

2. Natasha Saini, Nitin Pandey, Ajeet Pal Singh (2015)," Enhancement Of Security Using Cryptographic Techniques", 978-1-4673-7231-2/15©2015 *IEEE*.

3. Takahiro Fujita, Kiminao Kogiso, Kenji Sawada, & Seiichi Shin (2015) "Security Enhancements of Networked Control Systems Using RSA Public-Key Cryptosystem", 978-1-4799-7862-5/15©2015 *IEEE*.

4. Yasmin Alkady, Mohmed I. Habib, Rawya Y. Rizk (2013)," A New Security Protocol Using Hybrid Cryptography Algorithms", 978-1-4799-3370-9/13©2013 *IEEE*.

5. Bhushan Chaudhari, Prathmesh Gothankar, Abhishek Iyer, D. D. Ambawade (2012),"Wireless Network Security Using Dynamic Rule Generation of Firewall", *2012 International Conference on Communication, Information & Computing Technology (ICCICT)*, Oct. 19-20,2012.

6. Sangita A. Jaju, Santosh S. Chowhan (2015)" A Modified RSA Algorithm to Enhance Security for Digital Signature", 978- 1-4799-6908-1/15©2015 *IEEE*.

7. Ayman Tajeddine Ayman Kayssi Ali Chehab Imad Elhajj (2014)," Authentication Schemes for Wireless Sensor Networks", *17th IEEE Mediterranean Electrotechnical Conference*, Beirut, Lebanon, 13-16 April 2014. 978-1-4799-2337- 3/14©2014 IEEE.

8. Ashwak alabaichi, Adnan Ibrahem Salih, (2105)"Enhance Security of Advance Encryption Standard Algorithm Based on Key-dependent SBox", ISBN: 978-1-4673-6832-2©*2015 IEEE*.

9. Kyung-Ah Shim (2015)," A Survey of Public-Key Cryptographic Primitives in Wireless Sensor Networks" *IEEE Communications Survey & Tutorials*, Vol., No., 2012, 1553-877X (c) 2015 *IEEE*.

10. Madhumita Panda, Atul Nag,(2015) "Plain Text Encryption Using AES, DES & SALSA20 by Java Based Bouncy Castle API on Windows & Linux", 2015 *Second International Conference on Advances in Computing & Communication Engineering*, 978-1-4799-1734-1/15 © 2015 IEEE DOI 10.1109/ICACCE.2015.130.

11. S. Scott, R. Sylvia, K. Brad, et al(2003). Data-Centrie storage in sensor nets, *ACM SIGCOMM Computer*, Communication Review. 33(2003):137-142.

12. G. Abhishek, G. Jens, C. John (2003), Resilient data-centric storage in wireless ad-hoc sensor networks, *Proc. of the 4th Int'l Conf. on Mobile Data Management*, (2003):45-62.
13. Z. Wensheng, C. Guohong, L.P Tom (2003), Data dissemination with ring-based index for wireless sensor networks, *IEEE Int'l Conf. on Network Protocols*, (2003):305-314.
14. G. Benjamin, E. Deborah, G. Ramesh, et a (2003) l. DIFS: a distributed index for features in sensor networks, Proc. Of the 1st *IEEE Int'l Workshop on Sensor Network Protocols and Applications Anchorage.*(2003):163-173.
15. L. Xin, J.K Young, G. Ramesh, et al (2003). Multi-Dimensional range queries in sensor networks, Proc. of the 1st *Int'l Conf. on Embedded Networked Sensor Systems.*(2003):509-517.
16. R.H Wendi, C. Anantha, B. Hari (2000), Energy-Eficient communication protocol for wireless microsensor networks, *Proc. of the 33rd Hawaii Int'l Conf. on System Sciences.* (2000):8020-8029

Chapter 15

Diagnosis Of Various Diseases In Blood Using Different Sensors

Ramu Priya G
Department of ECE, Sethu Institute of Technology, Virudhu-
nagar, India

Abstract: Now a days, Stress is one of the main concerns when it comes to people's being and health. Knowing before which events may cause stress, can be helpful in managing it. This paper aims to provide a solution for predicting stress levels caused by future events. Stress care industry has perpetually been on the forefront in the adoption and utilization of information and communication technologies (ICT) for efficient stress care administration and treatment. Recent developments in ICT and the emergence of Internet of Things (IOT) have opened up new avenues for research and exploration in all fields including medical and stress care industry. Hospitals have started using the cell instruments for communication intent and for this intent internet of things (IOT) has been used and fused with wi-fi sensor node reminiscent of small sensor nodes. This system is used to measure the Blood flow, o2, co2, glucose, stress levels. Furthermore, we have performed our own validation experiment for the correlation of stress, which revealed a strong positive correlation. This offers us further confirmation that our system is capable of providing indicative estimations of stress levels for future events.

1. INTRODUCTION

In this busy life, people don't have enough time to visit a doctor for the routine check-up so the health issues go on increasing and people suffer from it. The Same scenario is faced by senior citizen's they cannot visit the hospitals regularly. People are also not ready to wait in the queue and appointments for the check- up. Sometimes if the person is suffering from a major health condition and the treatment is not available in the nearby locality so he has to travel all the way to the place where the treatment is available. With the help of the remote health monitoring system, you can check your health parameters by sitting at your home and you can also share these parameters with your doctor who is not in the nearby locality. If the patient is suffering from a major health issue and the present doctor is unable to help him, he can also show the parameters to the doctor who can help him in any condition. If the treatment is not available in your country you can also go for doctors in abroad and take suggestions from them. With the use of remote health monitoring system using Wi-Fi. The death rate due to simple health issues can be reduced and lifestyle of people can be improved. The remote health monitoring system is small and portable, the patient can take the device with himself whereverhe wants to. This device is a kind of one-time investment the servicing is very low and is very durable. If any major changes are identified, then it is notified.

Networked sensors, either worn on the body or embedded in our living environments, make possible the gathering of rich information indicative of our physical and mental health. Captured on a continual basis, aggregated, and effectively mined, such information can bring about a positive transformative change in the health care landscape. In particular, the availability of data at hitherto unimagined scales and temporal longitudes coupled with a new generation of intelligent processing algorithms can: (a) facilitate an evolution in the practice of medicine, from the current post facto diagnose-and treat reactive paradigm, to a proactive framework for prognosis of diseases at an incipient stage, coupled with prevention, cure, and overall management of health instead of disease, (b) enable personalization of treatment and management options targeted particularly to the specific circumstances and needs of the individual, and (c) help reduce the cost of health care while simultaneously improving outcomes. In this paper, we highlight the opportunities and challenges for IoT in realizing this vision of the future of health care.

The modern visionary of healthcare industry is to provide better healthcare to people anytime and anywhere in the world in a more economic and patient friendly manner. Therefore for increasing the patient care efficiency, there arises a need to improve the patient monitoring devices and make them more mobile. The medical world today faces two basic problems when it comes to patient monitoring.

Recent years have seen a rising interest in wearable sensors and today several devices are commercially available [1]–[3] for personal health care, fitness, and activity awareness. In addition to the niche recreational fitness arena catered to by current devices, researchers have also considered applications of such technologies in clinical applications in remote health monitoring systems for long term recording, management and clinical access to patient's physiological information [4].

A. Wearable Based Methods

Wearable based methods often rely on smart sensors with embedded processing. They can be attached to the human body or worn in their garments, clothing or jewelry. Zhang, Ren and Shi proposed HONEY (Home healthcare sentinel system), a three-step detection scheme which consisted of an accelerometer, audio, image andvideo clips. Its innovation was to detect falls by leveraging a triaxial accelerometer, speech recognition, and on-demand video. In HONEY, once the fall event was detected, an alert email was immediately sent and the fall video was uploaded to the network storage for further investigation. Bagalà *et al.* gave an evaluation of accelerometer- based fall detection algorithms on real-worldfalls.

B Vision Based Methods

Vision based methods are always related to spatiotemporal features, change of shape, and posture. Yu *et al.* proposed a vision based fall detection method by applying background subtraction to extract the foreground human body and post processing to improve the result. To detect a fall, information was fed into a directed acyclic graph support vector machine for posture recognition. This system reported a high fall detection rate and low false detection rate. Rougier analyzed human shape deformation during a video sequence which is used to track the person's silhouette.

C. Ambient Based Methods

Ambient based methods usually rely on pressure sensors, acoustic sensors or even passive infrared motion sensors, which are usually implemented around caretakers' houses The fall detection sensors are linear arrays of electret condensers placed on a pre-amplifier board. In order to capture the information of the sound height, the sensor array was placed in the z-axis. The limitation of this method was that that only one person was allowed in the vicinity. Winkley, Jiang and Jiang proposed Verity, a 2- component system which had

221

a based station and a direct monitoring device. In this particular system, ambient/skin temperatures were measured for real time monitoring.

BLOCK DIAGRAM:

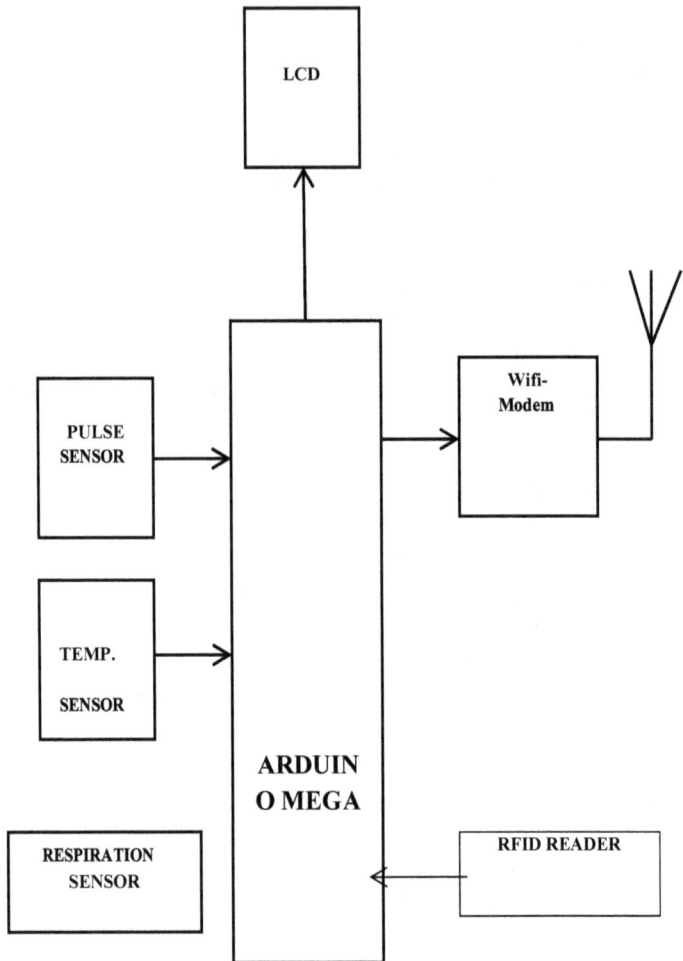

Complete Block Diagram

BLOCK DIAGRAM DESCRIPTION:

In today's world, the maximum use of resource is always compli-
mented. So, the use of wireless technology is enhanced to meet the
need of remote control and monitoring. Remote patient monitoring (RPM is a technology that enables us to monitor patient outside
of clinic or hospital without having to visit a patient. It may increase
access to health services and facilities while decreasing cost.

Remote Patient Monitoring saves time of both patient and doctor,
hence increasing efficiency and reliability of health services. Heartbeat and body temperature are the major signs that are routinely
measured by physicians after the arrival of a patient. Heart rate refers to how many times a heart contracts and relaxes in a unit of
time (usually per minute). Heart rate varies for different age groups.
For a human adult of age 18 or more years, a normal resting heart
rate is around 72 beats per minute (bpm). A lower heart rate at rest
implies more efficient heart function and better cardiovascular fitness. Like heart rate, normal body temperature also varies from
person to person and changes throughout the day.

The body temperature is lowest in the early morning and highest in
the early evening. The normal body temperature is about 37 °C or
98.6°. However, it can be as low as 36 .1 °C (97F) in the early morning and as high as 37.2 °C (99F) and still be considered normal.
Thus, the normal range for body temperature is 97 to 100 degrees
Fahrenheit or 36.1 to 37.8 degrees Celsius. The use of microcontroller is in every field even we can use it in the design and fabrication of
biomedical equipment. A little example is here. The microcontroller
at89s51 (8051) is here used to develop a heartbeat monitoring system. By placing your finger in between a LED and photo resistance,
She/he can detect the pulses of heart, the analog voltages are further processed with an operational amplifier LM 358, this chip has
two built in OPAMPs.[6] The TTL pulses or digital pulse are then
feed to the external interrupt of microcontroller 8051. By using a
softwarecounter in the code, they can count the pulses, and the result the process is displayed on an LCD (2 line 16 characters). HB =
5184/t, Where t is average of time delay between 2 consecutive
pulses here use of first 5 pulses for calculation of HB.

HARDWARE ARCHITECTURE:

Power Supply Design:

Every circuit needs a source to give energy to that circuit. The Source wills a particular voltage and load current ratings. The following is a circuit diagram of a power supply. We need a constant low voltage regulated power supply of +5V, providing input voltages to the microcontroller RS232, LM311 and LCD display which requires 5 volts supply.

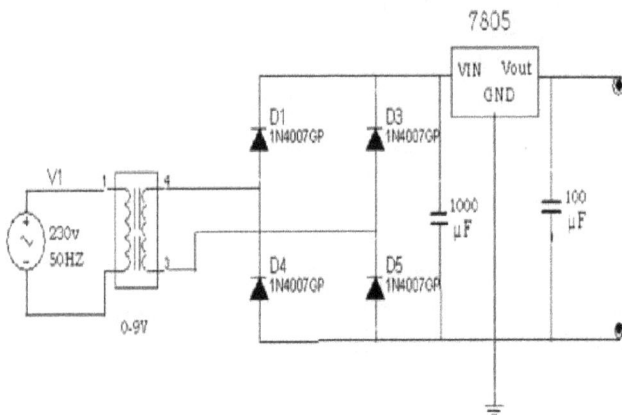

Schematic layout for Power Supply

Every power supply has the following parts:

- o Transformer
- o Rectifier
- o Capacitor (filter)
- o Regulator
- o Resistors

TRANSFORMER

WORKING PRINCIPLE OF TRANSFORMER:

The transformer works on the principle of faradays law of electro-magnetic inductions. Transformer in its simplest form.

The core is built up of thin laminations insulated from each other in order to reduce eddy current loss in the more. The winding are un-guarded from each other and also from thecare. The winding connected to the load is called the secondary winding for samplings they are shown on the opposite side of core but in practice they are distributed owner both sides of the cores.

Let us say that transformer has N1 turns in its primary winding and N2 turns in its secondary winding. The primary winding is connected to a sinusoidal voltage of magnitude V1 at a frequency FH2. A working flux is set up in magnetic core. The working flux is alternating and sinusoidal as the applied voltage is alternating and sinusoidal. When these flux link the primary and the secondary winding emf are induced in them.

In our electrical and electronic circuit we use two important components namely.

RESISTOR

CAPACITOR RESISTOR:

A resistor is an electric component. It has a known value of resistance. It is especially designed to introduce a desired amount of resistance in a circuit. A resistor is used either to control the flow of current or to produce a voltage drop. It is the most commonly used component in electrical and electronic circuits.

TYPES OF RESISTORS:

➢　　Carbon resistor

> Metal oxide resistor

> Metal film resistor

> Wire wound resistor

> Variable resistor-carbon resistor

CAPACITOR

Capacitor is an electrical device used for storing electrical energy. The stored electrical energy is the form of a current in to the circuits which the capacitor form a part. Capacitor is one of the important components used in Radio, TV and other electronic circuits.

TYPES OF CAPACITORS:

> Paper Capacitor

> Mica Capacitor

> Ceramic Capacitor

> Electrolytic Capacitor

> Variable Capacitor

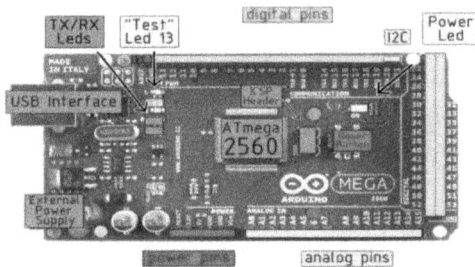

analog inputs and a larger space for your sketch it is the recommended board for 3D printers and robotics projects. This gives your projects plenty of room and opportunities maintaining the simplicity and effectiveness of the Arduino platform.

ARDUINO MEGA BOARDS

Mega 2560 is a microcontroller board based on the ATmega2560. It has 54 digital input/output pins (of which 15 can be used as PWM outputs), 16 analog inputs, 4 UARTs (hardware serial ports), a 16 MHz crystal oscillator, a USB connection, a power jack, an ICSP header, and a reset button. It contains everything needed to support the microcontroller; simply connect it to a computer with a USB cable or power it with a AC- to-DC adapter or battery to get started. The Mega 2560 board is compatible with most shields designed for the Uno and the former boards Duemilanove or Diecimila.

ARDUINO

The Arduino MEGA 2560 is designed for projects that require more I/O llines, more sketch memory and more RAM.

Technical specs

Microcontroller	ATmega2560
Operating Voltage	5V
Input Voltage (recommended)	7-12V
Input Voltage (limit)	6-20V
Digital I/O Pins	54 (of which 15 provide PWM output)
Analog Input Pins	16
DC Current per I/O Pin	20 mA
DC Current for 3.3V Pin	50 mA
Flash Memory	256 KB of which 8 KB used by bootloader
SRAM	8 KB
EEPROM	4 KB
Clock Speed	16 MHz
LED_BUILTIN	13
Length	101.52 mm
Width	53.3 mm
Weight	37 g

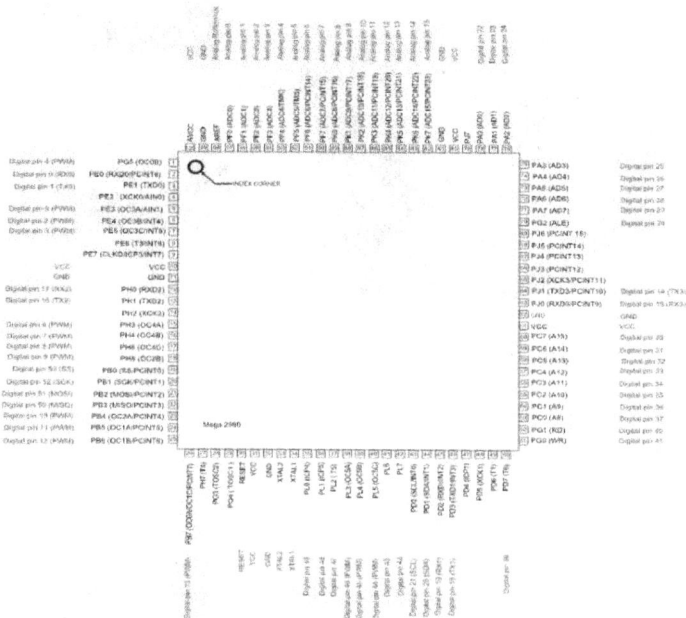

ARDUINO PIN LAYOUT

PULSE SENSOR

Pulse oximetry is a noninvasive method for monitoring a person's oxygen saturation (SO2). Its reading of SpO2 (peripheral oxygen saturation) is not always identical to the reading of SaO2 (arterial oxygen saturation) from arterial blood gas analysis, but the two are correlated well enough that the safe, convenient, noninvasive, inexpensive pulse oximetry method is valuablefor measuring oxygen saturation in clinical use.

In its most common (transmissive) application mode, a sensor device is placed on a thin part of the patient's body, usually a fingertip or earlobe, or in the case of an infant, across a foot. The device passes two wavelengths of light through the body part to a photodetector. It measures the changing absorbance at each of the wavelengths, allowing it to determine the absorbances due to the pulsing arterial blood alone, excluding venous blood, skin, bone, muscle, fat, and (in most cases) nail polish.

Sample output in processing

TEMPERATURE SENSOR:

LM35 is a IC temperature sensor with its output proportional to the temperature (in oC). The sensor circuitry is sealed and therefore it is not subjected to oxidation and other processes. With LM35, temperature can be measured more accurately than with a thermistor. It also possess low self heating and does not cause more than 0.1 oC temperature rise in precision still air.

Temperature Sensor layout

RESPIRATION SENSOR

The Respiration Sensor measures breathing rate and relative depth of abdominal or thoracic breathing. It is provided with an easy to apply elastic band and can be worn over clothing. The Respiration Sensor is usually placed over the abdomen. Respiration is often used in combination with the Blood Volume Pulse Sensor for HRV Training.

LIQUID CRYSTAL DISPLAY

LCD Diagram

An LCD is a small low cost display. It is easy to interface with a micro-controller because of an embedded controller(the black blob on the backof the board).This controller is standard across many displays which means many micro-controllers have libraries that make displaying messages as easy as a single line of code. LCDs with a small number of segments, such as those used in digital watches and pocket calculators, have individual electrical contacts for each segment. An external dedicated circuit supplies an electric charge to control each segment. This display structure is unwieldy for more than a few display elements.

INTERNET OF THINGS (IOT)

The internet of things, or IoT, is a system of interrelated computing devices, mechanical and digital machines, objects, animals or people that are provided with unique identifiers (UIDs) and the ability to transfer data over a network without requiring human-to-human or human-to-computer interaction.

A thing in the internet of things can be a person with a heart monitor implant, a farm animal with a biochip transponder, an automobile that has built-in sensors to alert the driver when tire pressure is low or any other natural or man-made object that can be assigned an IP address and is able to transfer data over a network.

Increasingly, organizations in a variety of industries are using IoT to operate more efficiently, better understand customers to deliver enhanced customer service, improve decision-making and increase the value of the business.

- ➤ monitor their overall business processes;

- ➤ improve the customer experience;

- ➤ save time and money;

- ➤ enhance employee productivity;

- ➤ integrate and adapt business models;

- ➤ make better business decisions; and

- ➤ generate more revenue. **SOFTWARE DESCRITION Arduino IDE**

Arduino is a flexible programmable hardware platform designed for artists, designers, tinkerers, and the makers of things. Arduino's little, blue circuit board, mythically taking its name from a local pub in Italy, has in a very short time motivated a new generation of DIYers of all ages to make all manner of wild projects found anywhere from the hallowed grounds of our universities to the scorching desert sands of a particularly infamous yearly arts festival and just about everywhere in between. Usually these Arduino based projects require little to no programming skills or knowledge of electronics theory, and more often than not, this handiness is simply picked up along the way.

ANDROID APPLICATION:

Android is an operating system based on the Linux kernel and designed primarily for touchscreen mobile devices such as smartphones and tablet computers. Initially developed by Android, Inc., which Googlebacked financially and later bought in 2001Android was unveiled in 2007 along with the founding of the Open Handset Alliance—a consortium of hardware, software, and telecommunication companies devoted to advancing

232

open standards for mobile devices. The first publicly available smartphone running Android, the HTC Dream, was released on October 22, 2008.

CONCLUSION AND FUTURE WORK

The project Remote viral fever recording system has been completed successfully and the output results are verified. The results are in line with the expected output. The project has been checked with both software and hardware testing tools. In this work LCD, Arduino controller, Temperature sensor,respiration sensor ,Pulse sensor, Relay, are chosen are proved to be more appropriate for the intended application. The project is having enough avenues for future enhancement

REFERENCES

[1]   Aleksandra C. Zoric, SinisaS.llic (2005), "PC Based Electrocardiography &Data Acquisition", *TELSIKS, IEEE*, pp 619-622, September 28- 30 2005.

[2]   Tia Gao, Dan Greenspan, Matt Welsh, Radford R. Juang, and Alex Alm (2012), "Real Time Patient Monitoring System Using Lab view", *International Journal of Scientific & Engineering Research*,April-2012.

[3]   Sherin Sebastian, Neethu Rachel Jacob (2012), "Remote Patient

Monitoring System Using Android Technology", *IJDPS,* September 2012.

[4]   C. Wen, M. Yen, K. Chang and R. Lee (2008), "Real-time ECG telemonitoring system design with mobile phone platform", *Measurement,* Volume 41, Issue 4, May 2008, Pages 463-470.

[5]   Wilkoff BL, Auricchio A, Brugada J, Cowie M, Ellenbogen KA, Gillis AM et al (2008). "HRS/EHRA Expert Consensus on the Monitoring of Cardiovascular Implantable Electronic Devices (CIEDs): Description of Techniques, Indications, Personnel, Frequency Ethical Considerations", *Euro pace* 2008;10:707–25.

[6]   Varma N, Epstein A, Schweikert R, Love C, Shah JA, Irimpen (2008) " Evaluation of efficacy and safety of remote monitoring for ICD follow-up" the TRUST trial, Circulation 2008, Vol. 118, No. 22, 2316, Abstract 4078.

[7]   Kiely DK (2000), "Resident characteristics associated with wandering in nursing home", *Int J Geriatric Psychiatry.* 2000

[8]   K. Lorincz et al.(2004), "Sensor Networks for Emergency Response: Challenges and Opportunities," *IEEE Pervasive Computing,* , October-December 2004.

[9]   J. Hill et al., "System Architecture Directions for Networked Sensors," in *Proc. 9th Int'l Conf. Architectural Support for Programming Languages and Operating Systems* (ASPLOS 2000)", ACM Press, pp.93-104,2000.

[10]   K. Lorincz and M. Welsh, A Robust, "Decentralized Approach to RF Based Location Tracking, tech. report TR- 19-04, Division of Eng. And Applied Sciences ", Harvard Univ., Cambridge, MA, 2004.

www.ingramcontent.com/pod-product-compliance
Lightning Source LLC
Chambersburg PA
CBHW050457190326
41458CB00005B/1318

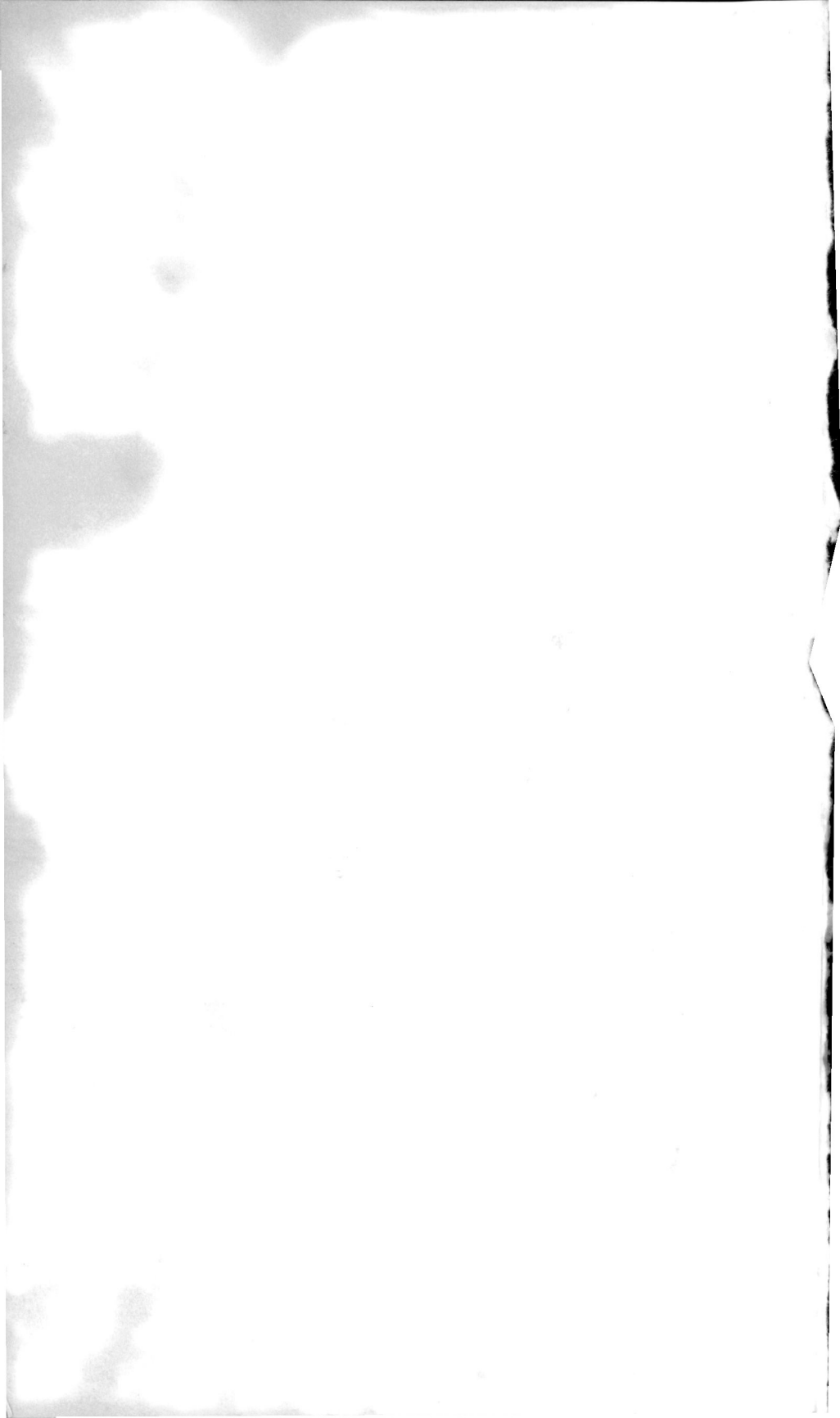